The Validation of Analytical Methods for Pharmaceutical Analysis

Mourne Training Services Ltd
5 Moor Hill Road, Newry, United Kingdom, BT34 2QJ
Tel: +44 (0)28 3083 4938
Email: info@mournetrainingservices.co.uk
Web: mournetrainingservices.com

ISBN 978-0-9561528-1-7

Preface

This book provides guidance on how to perform validation for the analytical methods which are used in pharmaceutical analysis. Validation of the analytical methods which are used during drug development and drug manufacturing is required to demonstrate that the methods are fit for their intended purpose. Additionally, the pharmaceutical industry around the world is subject to extensive regulations due to the nature of its products. Validation is a regulatory requirement and the data generated during an analytical method validation study is included in submissions to regulatory authorities for clinical trial and marketing applications.

The definitive reference for this topic is the guideline produced by the International Conference on Harmonisation of Technical Requirements for Registration of Pharmaceuticals for Human Use (ICH), Q2(R1) 'Validation of Analytical Procedures: Text and Methodology'. Sections of the guideline have been reproduced in this book with the kind permission of the ICH Secretariat. All ICH guidelines are available via the website, www.ich.org, and thus the validation guideline may be easily checked for revisions.

The guidance in this book is primarily aimed at analytical methods for small molecules. Reference is made to validation of methods for pharmaceuticals of biological origin, such as proteins and peptides. In principle the ICH guidelines should be applied to these types of compounds, however it is acknowledged they present particular challenges during validation, in particular relating to biological assays. The references provided in Chapter 1 provide more information on the validation of analytical methods for biopharmaceuticals/ biotechnology derived products.

At the back of the book there is a glossary to help the reader become familiar with the terminology used in analytical method validation. When a new term is introduced it is shown in **bold** to indicate to the reader that a definition is available in the glossary. In the appendix there is a list of abbreviations and also a question section so that the reader can test their understanding of the content.

Oona McPolin

Contents

Introduction

Validation is defined as 'finding or testing the truth of something'. When analytical methods are used to generate results about the characteristics of drug related samples it is vital that the results are trustworthy: they may be used as the basis for decisions relating to administering the drug to patients. A validation study is performed on an analytical method to ensure that reliable results are always obtained.

Validation in the Pharmaceutical Industry

Analytical method validation is just one type of validation required during drug development and manufacturing. To comply with the requirements of current Good Manufacturing Practices (GMP)[1-3] pharmaceutical companies should have an overall validation policy which documents how validation will be performed. This will include the validation of: production processes, cleaning procedures, analytical methods, in-process control test procedures, and computerised systems. The purpose of this validation is to show that processes involved in the development and manufacture of drugs, such as production, cleaning and analytical testing, can be performed in an effective and reproducible manner.

The reason that validation is included in cGMP in this way is to ensure that quality is built in at every step, and not just tested for at the end. 'Validation is intended to provide assurance of the quality of a system or process through a quality methodology for the design, manufacture and use of that system or process, that cannot be found by simple testing alone.'[4]

Data Quality

The quality of analytical data is assured by the combination of a number of critical components as shown in Figure 1. In the data quality triangle[5] the components are layered, each layer adds to the overall quality of the data. The base of the triangle is **Analytical Instrument Qualification** (AIQ), this is documented evidence that an instrument performs suitably for its intended purpose and that it is properly maintained and calibrated. The next layer is analytical method validation, the subject of this book. This is documented evidence that demonstrates that the analytical method is suitable for its intended use. The top layers of the triangle are **system**

suitability tests and **quality control checks**, these are used to demonstrate that the combination of system and method performed as expected at the time of the analysis. System suitability tests are commonly used for chemical analyses, which are usually subject to GMP regulations, and quality control checks are commonly used for bioanalytical analyses, which are usually subject to GLP regulations. Overall, AIQ and analytical method validation assure the quality before the analysis is performed and system suitability tests and quality control checks assure the quality immediately before or during the analysis.

Figure 1 Data quality triangle, from Basal et al in *AAPS PharmSciTech*[5]

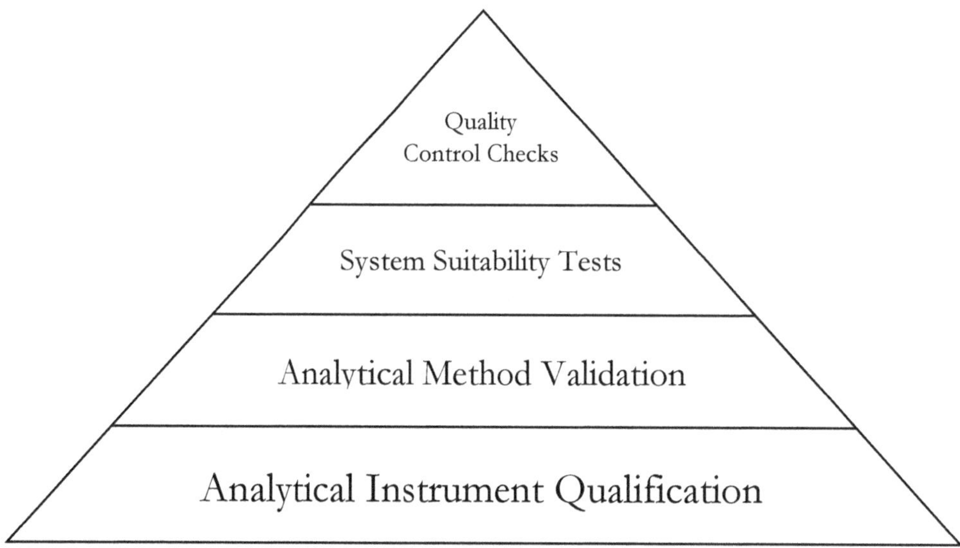

Analytical Instrument Qualification (AIQ)

The equipment which is used for all types of validation related to drug development and manufacture needs to be qualified. In the case of analytical methods this refers to the analytical instrumentation required to perform the test. Analytical method validation must be performed on appropriately qualified instruments.

The qualification procedure is usually carried out in four stages during which all actions are documented[5,6]. These are:

Design Qualification (DQ):
This covers all procedures prior to the installation of the system in the selected environment. The DQ defines the functional and operational specifications of the instrument and details the conscious decisions in the selection of the supplier. For commercial off the shelf (COTS) analytical instrumentation users generally have very little input into the design of the instrument and thus the DQ will detail the user requirements and the rationale for the selection of a particular supplier. For custom designed analytical instrumentation, the DQ details the key features of the design and how they address the user requirements.

Installation Qualification (IQ):
This covers all the procedures relating to the installation of the instrument in the selected environment. The IQ establishes that the instrument is received as designed and specified, that it is properly installed in the selected environment and that this environment is suitable for the operation and use of the instrument. The IQ may be carried out by the supplier and/or the user. For some complex instrumentation it may have to be performed by the supplier.

Operational Qualification (OQ):
This is the process of demonstrating that an instrument will function according to its operational specification in the selected environment. The OQ usually takes place after the IQ of a new instrument or after a change to the instrument, such as repair or change of location. As with the IQ, the OQ may be carried out by the supplier and/or the user. IQ/OQ is now offered by most suppliers of analytical instrumentation when a new instrument is purchased.

Performance Qualification (PQ):
This is defined as the process of demonstrating that an instrument consistently performs according to a specification appropriate for its routine use. PQ may be offered as part of a service contract for the routine maintenance of analytical instruments and/or may be carried out by the user. Calibration is part of performance qualification.

This summary gives an overview of equipment qualification, more information can be found in the GMP guidelines[1-3], also many articles and books are available on this topic[5-7].

The Purpose of an Analytical Method

An analytical method details the steps necessary to perform an analysis. This may include: preparation of samples, standards and reagents; use of the apparatus; generation of the calibration curve, use of the formulae for the calculation, etc. The objective of validation of an analytical method is to demonstrate that the method is suitable for the intended use.

The use of analytical methods during drug development and manufacturing provides information on:

- Potency, which can relate directly to the requirement of a known dose.
- Impurities, which can relate to the safety profile of the drug.
- Evaluation of key drug characteristics such as crystal form, drug release, and drug uniformity, properties which can compromise **bioavailability**.
- **Degradation products**, methods need to be **stability indicating**.

- Effect of key manufacturing parameters, to ensure that the production of drug substance and drug product is consistent.

The validation which is performed on the methods which generate this data needs to demonstrate that they can do so reliably and consistently.

The life cycle of an analytical method

Once a method has been developed and validated it may then be used for routine analysis, as shown in Figure 2. However, changes may occur which make it necessary to evaluate whether the method is still suitable for its intended use. The change may be covered by the existing validation, in which case no further validation is required or the change may result in revalidation, and in some cases, redevelopment of the method followed by validation of the new method.

Figure 2 The life cycle of an analytical method

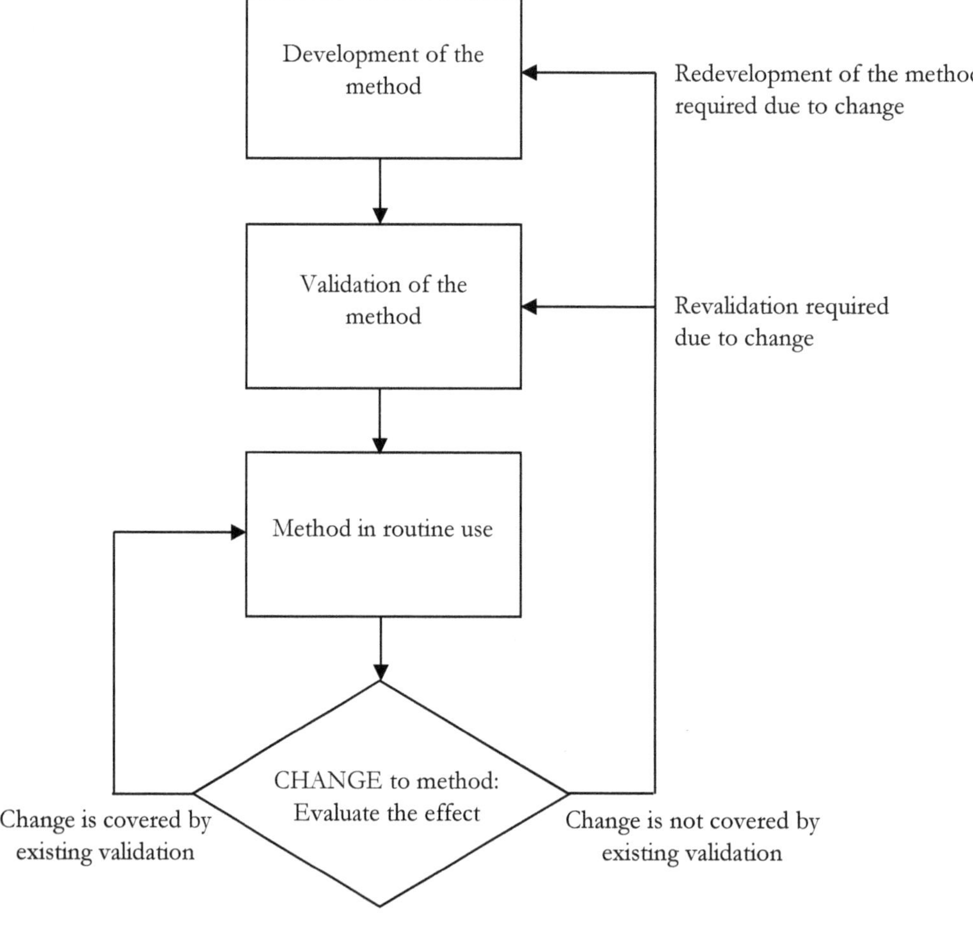

Examples of potential changes and the effect are shown in Table 1.

Table 1 Examples of potential changes to analytical methods

Source of Change	Change	Effect
Synthesis of the drug substance	Different synthetic route	Revalidation and potentially redevelopment of the method
Composition of the drug product	New excipient	Revalidation
	New dose size	Revalidation
Specification limits	Significantly lower impurity limits	Revalidation and potentially redevelopment of the method
	Significantly higher impurity limits within validation limits	None
New application of the method	New compound	Development and validation of new method

Validation Guidelines

The International Conference on Harmonisation of Technical Requirements for Registration of Pharmaceuticals for Human Use (ICH) brings together the regulatory authorities of Europe, Japan and the United States and experts from the pharmaceutical industry in the three regions to discuss scientific and technical aspects of product registration. Analytical method validation was one of the first topics within the quality section and resulted in a set of guidelines for method validation in pharmaceutical analysis.

The guidelines are referred to as '**Q2(R1): Validation of Analytical Procedures: Text and Methodology**'[8]. Prior to 2005 they existed as two documents 'Q2A: Text on Validation of Analytical Procedures', and 'Q2B: Validation of Analytical Procedures: Methodology'. The guidelines include a harmonised set of terms and definitions together with basic requirements for validation. The ICH guidelines provides a philosophical background to method validation rather than a checklist, 'It is the responsibility of the applicant to choose the validation procedure and protocol most suitable for their product'.[8]

Note: The term 'analytical procedure' as defined in the ICH guidelines is interchangeable with 'analytical method' or 'test procedure'.

The scope of the ICH guidelines is restricted to validation of analytical methods included as part of registration applications in the EC, Japan and USA, and does not necessarily seek to cover other areas of the world. In practice however, these guidelines are used internationally. There are also guidelines available from the US

Food and Drug Administration (FDA) which contain more detailed information on the US requirements. These are:

Draft guidance for industry: Analytical Procedures and Methods Validation[9]

This guideline has not yet been implemented but it provides useful information on FDA requirements for validation and characteristics which should be studied for different types of analytical methods.

Guidance for Industry: Bioanalytical Method Validation[10]

This guideline is for the validation of quantitative methods used for the determination of drugs and/or metabolites in biological matrices.

Reviewer Guidance: Validation of Chromatographic Methods[11]

This guideline has been written to provide information for FDA reviewers when evaluating validation of chromatographic methods. As such it provides useful information regarding FDA expectations and how validation data will be reviewed.

There are inconsistencies between some definitions of the criteria and the experimental procedures proposed to evaluate those criteria in the available guidelines on analytical method validation. Rozet et al. have conducted a review of these.[12]

Types of Analytical Methods

Universal tests

The ICH guideline[8] on performing analytical method validation is directed to the four most common types of methods. These are:

- Identification tests
- Quantitative tests for impurities
- Limit tests for the control of impurities
- Quantitative tests of the active moiety in samples of drug substance or drug product or other selected component(s) in the drug product.

The analytical tests of appearance, identification, assay and impurity determination are referred to as universal tests, i.e., tests which are considered to be potentially applicable to all new drug substances, or all new drug products. (Universal tests are defined in the ICH guideline on setting specifications for drug substances and drug products.[13]) Since appearance tests do not normally require validation, the ICH guidelines are applicable to the universal tests.

Identification

An identification method is intended to ensure the identity of an analyte in a sample, usually the **active pharmaceutical ingredient** in drug substance or drug product.

The test should be specific for the analyte, i.e., it should only give a positive response for that compound. However, specific methods may be difficult to achieve and in this case two independent methods of a nonspecific nature are normally used. A full discussion of specificity is included in Chapter 2. Tests which are used for identification include IR, differential scanning calorimetry (DSC), X-Ray diffraction (XRD), UV and HPLC retention time. Identification tests may be based on a comparison with a certified reference standard or on compliance with predetermined numerical limits. In the case where the drug exists in a salt form, a test for the counter-ion may also be required.

Assay

Assay tests quantify the analyte present in a given sample. For the drug substance this is the active pharmaceutical ingredient. In the case of the drug product this may be the active pharmaceutical ingredient(s) or other selected component(s), e.g., antimicrobial preservative content, antioxidant preservative content. A wide range of analytical techniques are used for assay methods including: HPLC, UV spectroscopy, titration and GC.

Impurities

The different types of impurities which are present in drug substance and drug products originate from a number of sources, see Table 2. The ICH guidelines[14-16] relating to impurities provides limits for the acceptable level of impurities in drug substances and drug products. Methods used to analyse impurities may be quantitative, where the amount of the impurity is quantified, or a limit test, which demonstrates that the level of the impurity is below a certain limit, may be applied.

Table 2 Classification of impurities in drug substance and drug product

Impurity Type	Examples	Typical origin
Organic	By-products Starting materials Intermediates Reagents	The synthetic process for the drug substance.
	Degradation Products	Degradation of the active pharmaceutical ingredient in both the drug substance and in the drug product.
	Impurities arising from excipients	Excipients in the drug product.
Inorganic	Reagents Catalysts Residual metals Inorganic salts Filter aids	The synthetic process for the drug substance including the equipment used and processing aids, such as filters.

Table 2 Classification of impurities in drug substance and drug product

Impurity Type	Examples	Typical origin
	Impurities arising from excipients	Excipients in the drug product.
Solvent	Reaction solvents Isolation solvents Chromatographic solvents	The synthetic process for the drug substance and also the manufacturing of the drug product.
Other	Polymorphic forms	Undesirable polymorphic forms of the active pharmaceutical ingredient.
	Enantiomeric impurities	The undesirable enantiomeric form of the active pharmaceutical ingredient.

The analysis of organic impurities is usually performed using a chromatographic technique, e.g., HPLC coupled with UV detection. Inorganic impurities are analysed using procedures found in **pharmacopoeias** and other standard references, e.g., residue-on-ignition test. Solvents are typically analysed using capillary GC. Examples of analytical techniques used for **polymorphic form** impurities include X-ray powder diffraction, Raman spectroscopy, and solid state NMR. Enantiomeric impurities are usually analysed using a chromatographic technique such as chiral HPLC or CE.

Analytical methods not included in the ICH guidelines

Although the ICH guidelines are limited to the universal tests of identification, assay and impurity determination, this does not mean that other types of methods do not need to be validated. All analytical methods should be validated unless the method employed is included in the relevant pharmacopoeia or other recognised standard reference. The principles discussed in the ICH guidelines apply to all types of analytical methods, the actual testing performed will depend on the nature of the method. However, specific recommendations in the available guidelines may not be applicable to certain unique methods for biological, biotechnological, botanical or radiopharmaceutical drugs.

Specific tests

'A specific test is a test which is considered to be applicable to particular new drug substances or particular new drug products depending on their specific properties and/or intended use.'[14]

Specific tests for drug substances which require validation include: Particle size; Polymorphic forms; Enantiomeric purity and Water content. Specific tests for drug products which require validation include: Dissolution; Uniformity of dosage units; Water content; Antimicrobial preservative content (oral liquids and parenterals); Antioxidant preservative content (oral liquids and parenterals); Extractables (oral liquids and parenterals); Alcohol content (oral liquids); Particle size (oral and injectable suspensions); Particulate matter (parenterals); Particle size distribution (suspensions); Osmolality (parenterals); Droplet distribution (inhalation dosage forms); and spray pattern (inhalation dosage forms).

Analytical methods relating to biotechnological/biological products

The ICH guideline[17] relating to specifications for biotechnological/biological products (applies to proteins and polypeptides, their derivatives, and products of which they are components) states that validation of the analytical procedures applied to these products should be performed in accordance with the ICH guideline[8] on validation of analytical procedures, 'except where there are specific issues for unique tests used for analyzing biotechnological and biological products'.[17] These unique tests need to be considered on a case by case basis and the most appropriate approach selected on a scientific basis. Due to the nature of these products, the validation of the analytical methods used for testing may present a challenge. There are a number of articles about this topic in the literature.[18,19,20] Supplementary information regarding validation of methods which are applied to biotechnology derived pharmaceuticals is also available in the United States Pharmacopeia.[21]

Bioanalytical methods

These are defined as methods 'performed for the quantitative determination of drugs and/or metabolites in biological matrices such as blood, serum, plasma or urine.'[10] They are commonly used in human clinical **pharmacology**, bioavailability (BA), and **bioequivalence** (BE) studies requiring **pharmacokinetic** (PK) evaluation. They are also used for non-human pharmacology/**toxicology** studies and preclinical studies. Comprehensive guidelines are available from the FDA[10] to assist with the validation of these methods and thus show that they are reliable and reproducible for the intended use. There are a number of articles concerning the interpretation of these guidelines.[22,23,24]

Microbiological methods

Microbiological tests required for drug substance and drug product are validated in accordance with the relevant procedures detailed in the pharmacopoeias. When performing the validation consideration must be given to the unique nature of microbiological assays.[25]

Other analytical methods

Analytical methods which are used in drug development and manufacture but do not analyse the drug substance or drug product are not covered by the available validation guidelines. This includes methods to support cleaning validation and analysis of environmental samples, and also methods for **excipients**, raw materials and intermediates. Similar principles to those used for drug substances and drug products may be applied to achieve validation of these methods.

Analytical techniques

The type of analytical technique used by an analytical method influences the way in which the validation investigation is carried out. The ICH guidelines contain several recommendations for how validation should be applied to chromatographic techniques. For other techniques a sound understanding of the method, combined with a scientific approach, is required to evaluate the most suitable way to perform validation. An example of a technique where a slightly different approach is required is near-infrared (NIR) analysis. Guidelines to assist in the application of the ICH concepts for the development and validation of NIR methods are available.[26,27]

Validation Characteristics

Validation of an analytical method is achieved by performing testing on a number of validation characteristics. These are listed below together with the definition given in the ICH guidelines.[8]

Specificity

'Specificity is the ability to assess unequivocally the analyte in the presence of components which may be expected to be present. Typically, these might include impurities, degradants, matrix, etc.'

Accuracy

'The accuracy of an analytical procedure expresses the closeness of agreement between the value which is accepted either as a conventional true value or an accepted reference value and the value found. This is sometimes termed trueness.'

Precision

'The precision of an analytical procedure expresses the closeness of agreement (degree of scatter) between a series of measurements obtained from multiple sampling of the same homogeneous sample under the prescribed conditions.

The precision of an analytical procedure is usually expressed as the variance, standard deviation or coefficient of variation of a series of measurements.

Repeatability

Repeatability expresses the precision under the same operating conditions over a short interval of time. Repeatability is also termed intra-assay precision.

Intermediate precision

Intermediate precision expresses within-laboratories variations: different days, different analysts, different equipment, etc.

Reproducibility

Reproducibility expresses the precision between laboratories (collaborative studies, usually applied to standardization of methodology).'

Detection Limit

'The detection limit of an individual analytical procedure is the lowest amount of analyte in a sample which can be detected but not necessarily quantitated as an exact value.'

Quantitation Limit

'The quantitation limit of an individual analytical procedure is the lowest amount of analyte in a sample which can be quantitatively determined with suitable precision and accuracy. The quantitation limit is a parameter of quantitative assays for low levels of compounds in sample matrices, and is used particularly for the determination of impurities and/or degradation products.'

Linearity

'The linearity of an analytical procedure is its ability (within a given range) to obtain test results which are directly proportional to the concentration (amount) of analyte in the sample.'

Range

'The range of an analytical procedure is the interval between the upper and lower concentration (amounts) of analyte in the sample (including these concentrations) for which it has been demonstrated that the analytical procedure has a suitable level of precision, accuracy and linearity.'

Robustness

'The robustness of an analytical procedure is a measure of its capacity to remain unaffected by small, but deliberate variations in method parameters and provides an indication of its reliability during normal usage.'

Summary

1. Validation is an important procedure in the pharmaceutical industry and is used to ensure that quality is built into the processes supporting drug development and manufacture.
2. Data quality is assured by the combination of four components: Analytical Instrument Qualification (AIQ); analytical method validation; system suitability tests and quality control checks.
3. Validation of an analytical method is intended to demonstrate that it is suitable for its intended use.
4. Guidelines for analytical method validation as applied to pharmaceuticals are available from the International Conference on Harmonisation (ICH) of Technical Requirements for Registration of Pharmaceuticals for Human Use.
5. The type of method and analytical technique used will determine the nature and extent of the validation studies required. The most common methods for validation are identification, assay and impurities determination.
6. The characteristics of validation are: Specificity, linearity, range, accuracy, precision, detection limit, quantitation limit and robustness.

References

1. International Conference on Harmonisation (ICH) of Technical Requirements for Registration of Pharmaceuticals for Human Use, Topic Q7: **Good Manufacturing Practices for Pharmaceutical Ingredients**, 2000, www.ich.org.

2. **Current Good Manufacturing Practices for finished pharmaceuticals**, 21 CFR, Parts 210 and 211, US Food and Drug Administration, www.fda.gov.

3. EUDRALEX: The Rules Governing Medicinal Products in the European Community, Volume IV, **Good Manufacturing Practice for Medicinal Products, Annex 15, Qualification and Validation**, 2001.

4. R. D. McDowall, *Qual Assur J*, **9**, 3, 196-227, 2005, '**Effective and practical risk management options for computerised system validation**'.

5. S. K. Basal, T. Layloff, E. D. Bush, M. Hamilton, E. A. Hankinson, J. S. Landy, S. Lowes, M. M. Nasr, P. A. St. Jean, V. P. Shah, *AAPS PharmSciTech* 5 (1), Article 22, 2004, '**Qualification of Analytical Instruments for Use in the Pharmaceutical Industry: a Scientific Approach**'.

6. P. Bedson, M. Sargent, *Accred. Qual. Assurance* **1**, 265-274, 1996, '**The development and application of guidance on equipment qualification of analytical instruments**'.

7. L. Huber, '**Validation and Qualification in Analytical Laboratories**', 2nd ed., CRC Press Inc, 2007.

8. International Conference on Harmonisation (ICH) of Technical Requirements for Registration of Pharmaceuticals for Human Use, Topic Q2 (R1): **Validation of Analytical Procedures: Text and Methodology**, 2005, www.ich.org.

9. Draft guidance for industry: **Analytical Procedures and Methods Validation**, US Food and Drug Administration, Center for Drugs and Biologics, Department of Health and Human Services, 2000, www.fda.gov.

10. Guidance for Industry: **Bioanalytical Method Validation**, US Food and Drug Administration, Center for Drugs and Biologics, Department of Health and Human Services, 2001, www.fda.gov.

11. Reviewer Guidance: **Validation of Chromatographic Methods**, US Food and Drug Administration, Center for Drugs and Biologics, Department of Health and Human Services, 1994, www.fda.gov.

12. E. Rozet, A. Ceccato, C. Hubert, E. Ziemons, R. Oprean, S. Rudaz, B. Boulanger, P. Hubert, *J. Chromatogr.* **A1158**, 111-125, 2007, '**Analysis of recent pharmaceutical regulatory documents on analytical method validation**'.

13. International Conference on Harmonisation (ICH) of Technical Requirements for Registration of Pharmaceuticals for Human Use, Topic Q6A: **Specifications: Test Procedures and Acceptance Criteria for New Drug Substances and New Drug Products**, 1999, www.ich.org.

14. International Conference on Harmonisation (ICH) of Technical Requirements for Registration of Pharmaceuticals for Human Use, Topic Q3A(R2): **Impurities in New Drug Substances**, 2006, www.ich.org.

15. International Conference on Harmonisation (ICH) of Technical Requirements for Registration of Pharmaceuticals for Human Use, Topic Q3B(R2): **Impurities in New Drug Products**, 2006, www.ich.org.

16. International Conference on Harmonisation (ICH) of Technical Requirements for Registration of Pharmaceuticals for Human Use, Topic Q3C(R3): **Impurities: Guidelines for Residual Solvents**, 2002, www.ich.org.

17. International Conference on Harmonisation (ICH) of Technical Requirements for Registration of Pharmaceuticals for Human Use, Topic Q6B: **Specifications: Test Procedures and Acceptance Criteria for Biotechnological/Biological Products**, 1999, www.ich.org.

18. S. O. Krause, *BioPharm International*, **17** (10), 52-61, 2004, '**Development and validation of analytical methods for biopharmaceuticals, part I: development and optimization**'.

19. S. O. Krause, *BioPharm International*, **17** (11), 46-52, 2004, '**Development and validation of analytical methods for biopharmaceuticals, part II: formal validation**'.

20. S. O. Krause, *BioPharm International*, **18** (3), 2005, '**Analytical Method Validation for Biopharmaceuticals: A Practical Guide**'.

21. United States Pharmacopeia (USP) 30, Biotechnology-derived articles – Tests <1047>.

22. S. Bansal, A.Destefano, *AAPS J.*, **9** (1), E109-114, 2007, '**Key elements of bioanalytical method validation for small molecules**'.

23. M. Kelley, B. DeSilva, *AAPS J.*, **9** (2), E156-163, 2007, '**Key elements of bioanalytical method validation for macromolecules**'.

24. J. W. Findlay, W. C. Smith, J. W. Lee, G. D. Nordblom, I. Das, B. S. DeSilva, M. N. Khan, R. R. Bowsher, *J. Pharm. Biomed. Anal.*, **21** (6), 1249-1273, 2000, '**Validation of immunoassays for bioanalysis: a pharmaceutical industry perspective**'.

25. Parenteral Drug Association, *PDA Journal of Pharmaceutical Science and Technology,* 54 (Suppl. TR33), 2000, PDA Technical Report No. 33: '**Evaluation, validation and implementation of new microbiological testing methods**', www.pda.org.

26. N. Broad, P. Graham, P. Hailey, A. Hardy, S. Holland, S. Hughes, D. Lee, K. Prebble, N. Salton, P. Warren, '**Guidelines for the Development and Validation of Near-infrared Spectroscopic Methods in the Pharmaceutical industry**' in 'Handbook of Vibrational Spectroscopy', Eds. J. M. Chalmers, P. R. Griffiths Wiley, 2002.

27. CPMP/QWP/3309/01 and EMEA/CVMP/961/01, '**Note for guidance on the use of near infrared spectroscopy by the pharmaceutical industry and the data requirements for new submissions and variations**', EMEA, London, 2003 (note that this guideline is currently under revision).

Further reading

J. Ermer, J. H. Miller, '**Method Validation in Pharmaceutical Analysis, a Guide to Best Practice**', Wiley, 2005.

D. M. Bliesner, '**Validating Chromatographic Methods: A practical guide**', Wiley, 2006.

C. C. Chan, Y. C. Lee, H. Lam, X. Zhang, '**Analytical Method Validation and Instrument Performance Verification**', Wiley, 2004.

C. M. Riley, T. W. Rosanske, eds., '**Development and Validation of Analytical Methods**', Pergamon, 1996.

Notes

Validation Characteristics

Validation of an analytical method is performed by consideration of a number of characteristics as introduced in Chapter 1. The nature of the analytical method will determine which characteristics should be investigated. In this chapter each characteristic is considered separately in the order in which they are listed in the ICH guidelines[1]. The guidance for each as written in the ICH guidelines is provided together with a discussion of the interpretation of the guidance and of the associated practical considerations.

Validation during Method Development

When validation is performed for analytical methods which are used to analyse drug related samples the nature of the investigation is usually confirmatory. During the development of the method many of the validation characteristics have already been considered although probably not in a formal way. For example:

- The concentration of the samples used for the analysis will have been selected so as to achieve a suitable response, e.g., in the case of a HPLC method to determine impurities using a UV detector, the concentration will be selected to fall within the linear range of the detector and to have a suitable response for the low concentrations required.

- The specificity will have been considered during the initial method development, e.g., in any chromatographic method the separation of the components of interest is the basis of the method development process and thus the retention time of components of interest and potential interferences are evaluated during the development of the method.

- An evaluation of the robustness of the method is usually commenced as soon as the method is developed. This will include identification of the critical parameters of the method and an evaluation of the stability of test solutions.

The result of this information gained during method development is that the validation experiments are expected to comply with the selected acceptance criteria and major problems are not usually anticipated. However, it is always possible that the validation study may identify problems that could result in changes to the method.

Statistics in Analytical Method Validation

Statistical analysis of the data produced during a method validation study is required to demonstrate the validity of the method. The statistical procedures which are commonly used are described during the discussion of each validation characteristic in this chapter and also in Chapter 4. In this preceding section the use of the mean, standard deviation and confidence intervals are considered.

Mean

The mean of a dataset is the total sum of all the values in the dataset divided by the number of values in the dataset. It is denoted by $\bar{x}$ (x bar) and is calculated using the following equation:

$$\bar{x} = \frac{1}{n}\sum x_i$$

Consider as an example the analysis of a pharmaceutical sample to determine the amount of active pharmaceutical ingredient present. A single determination or measurement will provide an estimation of the true value. If another determination is performed, then the result is likely to be different from the first determination but which is the best estimate of the true value? The mean of the results obtained for the two determinations is normally calculated to provide a better estimate for the true value. As the number of determinations increases the calculated sample mean will be nearer to the true value. In this situation the true value is the population mean, denoted by μ.

Standard deviation

The standard deviation of a dataset is a measure of the spread of the values in the dataset. It is calculated by measuring the difference between the mean and the individual values in the dataset. The standard deviation for a sample is determined using the following equation:

Standard deviation, $s = \sqrt{s^2}$

Where:

$$s^2 = \frac{1}{n-1}\sum_i \left(x_i - \bar{x}\right)^2$$

The calculation for standard deviation assumes a normal distribution of the data. This distribution is shown in Figure 3 and is typical for the results of most physico-chemical analysis. The data is centred about the mean with the majority of the observations near to the mean value. Fewer and fewer observations occur the further they are away from the mean. 68.3 % of the observations lie within 1 standard deviation of the mean, 95.5 % of the observations lie within two standard deviations and 99.7 % lie within 3 standard deviations. When using small data sets, typical in pharmaceutical analysis, a normal distribution may not be apparent on inspection of the data but this distribution can usually be assumed.

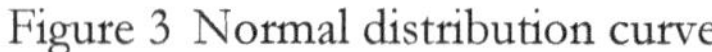
Figure 3 Normal distribution curve

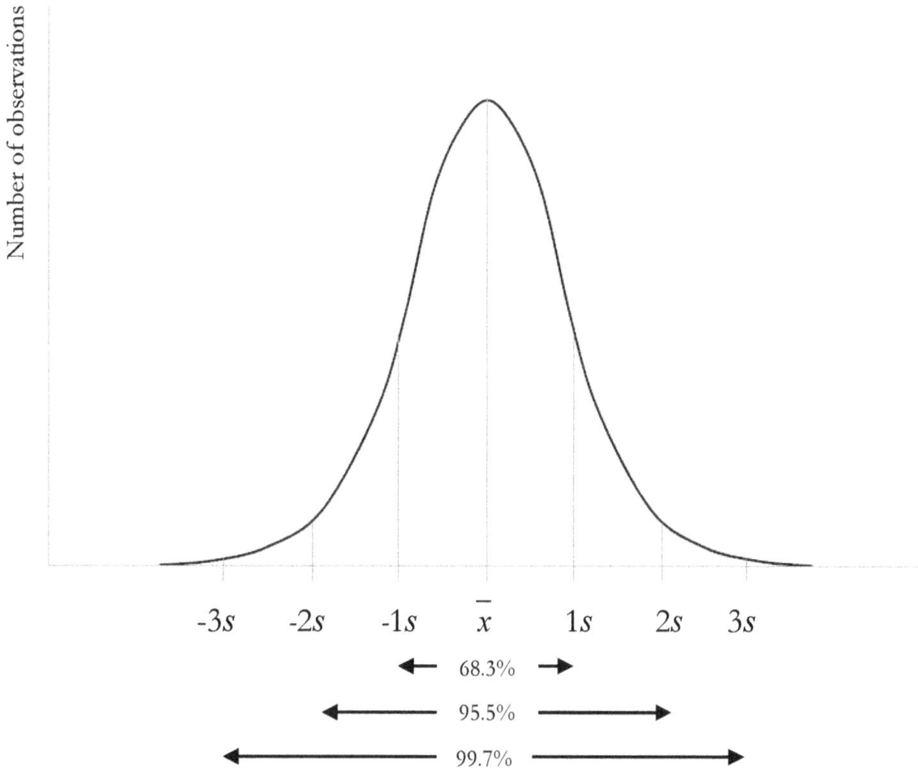

If the population standard deviation, σ, is defined as the standard deviation calculated for a large data set, and the sample standard deviation, s, is defined as the standard deviation of a small subset of this large data set, then it is found that the value calculated for the sample standard deviation on a number of subsets may vary considerably due to random variability in the data. The smaller the number of data, the higher is the variability of the sample standard deviation. As a result of this, it is recommended that the standard deviation should only be calculated for data sets where $n>3$.

Confidence intervals

Confidence intervals are used to indicate the reliability of an estimate. In the example quoted previously, where the amount of active pharmaceutical ingredient present in a pharmaceutical sample is analysed, the mean result obtained is an estimate of the actual amount present. A confidence interval provides limits around the experimentally determined value of the mean within which the true value (or population mean, μ) lies with a given degree of probability, usually 95%.

The size of the confidence interval depends on the value of the standard deviation, s. The confidence interval determined for a sample where the standard deviation approximates the population standard deviation can be significantly lower than if the estimate is based on a small dataset. If there is reason to believe that s, the sample standard deviation, is a good approximation of σ, the population standard deviation, then the following equation may be used to calculate the confidence interval:

$$\text{Confidence interval for } \mu = \bar{x} \pm \frac{z\sigma}{\sqrt{n}}$$

Where z depends on the confidence level required, see Table 3, and n is the number of values in the dataset.

Table 3 Confidence levels for various values of z

Confidence Levels, %	z
50	0.67
68	1.00
80	1.29
90	1.64
95	1.96
99	2.58
99.9	3.29

It can be seen that the values of z for the various confidence levels in Table 3 correspond to the normal distribution curve in Figure 3.

During analytical method validation studies, the calculation of the standard deviation is often based on a small dataset and an estimation of the population standard deviation, σ, is not possible. Thus, the confidence interval is broader than if σ is known. The variability in the standard deviation resulting from small datasets may be accounted for using the Student's t-distribution, refer to Figure 4. This is a probability distribution used when the sample size is small. The t-distribution is symmetrical and bell shaped, similar to the normal distribution, but there is a different t-distribution for each sample size, therefore the degrees of freedom is required to specify a specific t-distribution. The 'degrees of freedom' is an expression used to indicate the number of independent values available to estimate another value. In this case the degrees of freedom equals n – 1, where n is the number of values in the dataset. The spread of the t-distribution is greater than that of the normal distribution but decreases as the degrees of freedom increases.

Figure 4 Student's t-distribution

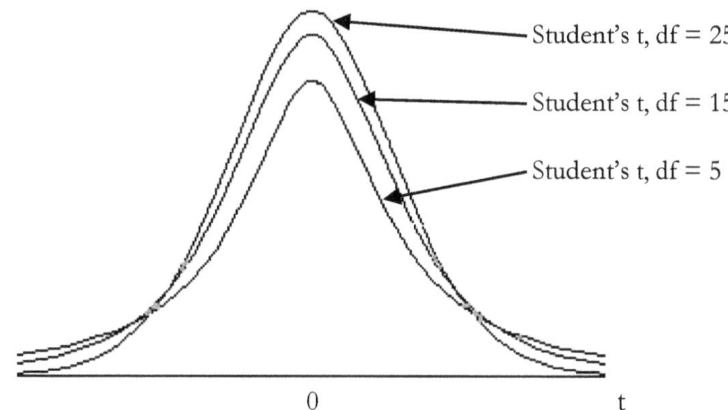

The following equation may be used to calculate the confidence interval:

$$\text{Confidence interval for } \mu = \bar{x} \pm \frac{ts}{\sqrt{n}}$$

Where t depends on the confidence level required and the degrees of freedom (n - 1), see Table 4, and n is the number of values in the dataset.

Table 4 Values of t for various levels of probability

Degrees of Freedom	Factor for Confidence Interval		
	90%	95%	99%
1	6.31	12.7	63.7
2	2.92	4.30	9.92
3	2.35	3.18	5.84
4	2.13	2.78	4.60
5	2.02	2.57	4.03
6	1.94	2.45	3.71
∞	1.64	1.96	2.58

See the further reading section at the end of this chapter for references relating to the use of statistics.

Specificity

The section on specificity in the ICH guidelines[1] is as follows:

ICH

1. Specificity

Specificity is the ability to assess unequivocally the analyte in the presence of components which may be expected to be present. Typically these might include impurities, degradants, matrix, etc. Lack of specificity of an individual analytical procedure may be compensated by other supporting analytical procedures(s).

An investigation of specificity should be conducted during the validation of identification tests, the determination of impurities and the assay. The procedures used to demonstrate specificity will depend on the intended objective of the analytical procedure.

It is not always possible to demonstrate that an analytical procedure is specific for a particular analyte (complete discrimination). In this case a combination of two or more analytical procedures is recommended to achieve the necessary level of discrimination.

1.1 Identification

Suitable identification tests should be able to discriminate between compounds of closely related structures which are likely to be present. The discrimination of a procedure may be confirmed by obtaining positive results (perhaps by comparison with a known reference material) from samples containing the analyte, coupled with negative results from samples which do not contain the analyte. In addition, the identification test may be applied to materials structurally similar to or closely related to the analyte to confirm that a positive response is not obtained. The choice of such potentially interfering materials should be based on sound scientific judgement with a consideration of the interferences that could occur.

1.2 Assay and Impurity Test(s)

For chromatographic procedures, representative chromatograms should be used to demonstrate specificity and individual components should be appropriately labelled. Similar considerations should be given to other separation techniques.

Critical separations in chromatography should be investigated at an appropriate level. For critical separations, specificity can be demonstrated by the resolution of the two components which elute closest to each other.

In cases where a non-specific assay is used, other supporting analytical procedures should be used to demonstrate overall specificity. For example, where a titration is adopted to assay the drug substance for release, the combination of the assay and a suitable test for impurities can be used.

The approach is similar for both assay and impurity tests:

1.2.1 Impurities are available

For the assay, this should involve demonstration of the discrimination of the analyte in the presence of impurities and/or excipients; practically, this can be done by spiking pure substances (drug substance or drug product) with

appropriate levels of impurities and/or excipients and demonstrating that the assay result is unaffected by the presence of these materials (by comparison with the assay result obtained on unspiked samples).

For the impurity test, the discrimination may be established by spiking drug substance or drug product with appropriate levels of impurities and demonstrating the separation of these impurities individually and/or from other components in the sample matrix.

1.2.2 Impurities are not available

If impurity or degradation product standards are unavailable, specificity may be demonstrated by comparing the test results of samples containing impurities or degradation products to a second well-characterized procedure e.g.,: pharmacopoeial method or other validated analytical procedure (independent procedure). As appropriate, this should include samples stored under relevant stress conditions: light, heat, humidity, acid/base hydrolysis and oxidation.

- for the assay, the two results should be compared;

- for the impurity tests, the impurity profiles should be compared.

Peak purity tests may be useful to show that the analyte chromatographic peak is not attributable to more than one component (e.g., diode array, mass spectrometry).

Specificity and selectivity

The definition of specificity given in the ICH guidelines is commonly used in the context of the validation of analytical methods for pharmaceuticals. However, it is important to note that in the field of analytical chemistry as a whole this definition is applied to the term **selectivity**. A distinction between these terms may be made as follows, 'a specific reaction or test is one that occurs only with the substance of interest, while a selective reaction is one that can occur with other substances but exhibits a degree of preference for the substance of interest. Few reactions are specific, but many exhibit selectivity.'[2] As a consequence of this the use of the term specificity is discouraged in the greater field of analytical chemistry[3] but persists in some areas of pharmaceutical analysis at present due to its use in the ICH guidelines. (Note that the FDA guidelines[4] for bioanalytical method validation refer to selectivity only.)

Identification methods

Since the purpose of an identification method is to ensure the identity of an analyte in a sample of drug substance or drug product then the method needs to be able to discriminate between the sample of interest and any other sample that may be mistaken for it. Examples of potential samples are: starting materials; isolated intermediates; other materials manufactured in the same plant; excipients in the drug product; enantiomers; solvates; and polymorphs. To demonstrate the specificity of the method it is applied to these compounds to show that a positive response is not obtained. For an identification method which is used to test for a counter-ion, other counter-ions would be tested using the method. The selection of suitable compounds for a specificity study requires a sound scientific understanding of both the method being used and the manufacturing process of the drug substance or drug product.

Assay and impurity methods

Some assay methods are non-specific, for example an assay which is based on a titration. This means that the impurities present in the sample may also contribute to the assay result. Assay methods of this type need to be supported by validated impurity methods which demonstrate that the effect of the impurities present is not significant.

To ensure that a correct result is calculated for an assay method it is important that there is no interference with the response due to the analyte. For an impurity method it is important that there is no interference with the response due to the active pharmaceutical ingredient and also with the response due to the impurities of interest. Although the requirements are slightly different the approach used to demonstrate specificity is similar for both. In both cases examples of potentially interfering compounds are impurities (examples of the different types of impurities which may be present in drug samples were listed in Table 2), and excipients. To demonstrate specificity, the method must be performed in the presence of the potentially interfering compounds to show that there is no interference.

Impurities are available

Samples of excipients are usually easily available, however samples of impurities may be difficult to obtain. This will depend on the stage of development of the drug and thus how much is known about the impurities generated during the synthetic process and also the degradation of the drug on storage. For drugs which are in the late stages of development, or already approved and being manufactured, the impurity profile of the drug is usually well known and samples of impurities, synthetic and degradation products, may have been prepared for characterisation. In this case section 1.2.1 (impurities are available) in the ICH guidelines may be followed.

It is important to consider the level at which an impurity may be present in a typical sample. For example, an impurity in a chromatographic method may be fully resolved from the peak of interest when it is present at a typical concentration. If it is evaluated for specificity at a high concentration the resulting larger peak may not be resolved from the peak of interest leading to an erroneous conclusion that the method is not specific. Therefore potential interferences should be investigated at the level at which they would be expected.

Impurities are not available

For drugs in earlier stages of development the impurity profile is usually still under investigation and samples of impurities are often not available. In addition, assay and impurity methods need to be stability indicating. This means that they must be suitable for degradation products which are generated during storage. The challenge of developing and validating a method which can analyse impurities that are not yet known is usually addressed by use of a **forced degradation** study. Forced degradation studies are typically used to develop suitable stability indicating assay and impurity methods and thus the samples generated may be available to demonstrate specificity during validation of the method.

Drug substance or drug product is degraded under a number of conditions to gain information on what degradation products might be expected. In section 1.2.2 of the ICH guidelines (impurities are not available) stress conditions of light, heat, humidity, acid/base hydrolysis, and oxidation are suggested. Further information on stress testing is provided in the ICH guidance relating to stability testing [5,6] and in a paper by Reynolds et al.[7] When using stress studies it is important to ensure that the degradation achieved is realistic and not excessive, typically degradation of the active ingredient by approximately 10% is considered sufficient.

In section 1.2.2 it is suggested that stressed samples are analysed by an additional validated procedure and the assay results or impurity profile compared. In practice it is unusual to have an additional method available for this comparison.

Chromatographic techniques

Chromatographic techniques are commonly used for the determination of impurities and thus statements and guidance which are particular to chromatographic techniques are included in the ICH guidelines. The use of **peak purity** testing is recommended to show that the analyte chromatographic peak is not attributable to more than one component. The methods of diode array and mass spectrometry are suggested. Neither method can prove that the peak observed is due to one component, but an investigation can give confidence in the method. The purity of the chromatographic peak is an integral requirement for specificity but unfortunately there are limitations to the available methods of peak purity determination.

Diode array detection generates a UV spectrum for every point on a chromatogram and therefore any changes in the UV spectra over the time it takes to elute a peak can be monitored. If two components which have different UV spectra are eluting together then a change may be observed or the UV spectrum will be different to that of a certified reference standard analysed in the same way. However, it is very common that impurities of drug components are closely related in structure to the drug itself and thus the UV spectrum for each is very similar if not identical. Therefore diode array cannot detect that there are two components present. Also impurities that are present at levels less than 1% of the active component are difficult to detect.

Mass spectrometry is a more powerful technique for peak purity. Mass spectra are obtained over the time taken to elute the peak. Additional masses may indicate the presence of a co-eluting impurity and single ion chromatograms for the extracted masses may be obtained and compared. Detection will depend on the individual MS response of the impurities. However, if the impurity has the same mass as the active component, then it will not be detected.

Another approach for the investigation of peak purity is to collect the fraction of mobile phase containing the peak of interest (i.e., preparative HPLC) and subject it to further analysis using spectroscopic techniques or alternative chromatographic methods.

Bioanalytical methods

Examples of potential interferences which may be present in bioanalytical samples are biological matrix (plasma, urine, or other matrix), **metabolites**, and decomposition products. All potential interferences should be tested to demonstrate non-interference and the FDA guidelines on bioanalytical method validation[4] recommend testing of the matrix blank from at least six sources. In addition, these guidelines make recommendations on selectivity issues particular to microbiological and **ligand-binding assays**.

Other methods

The demonstration of specificity for other methods will depend on the nature of the method and should be assessed on a case by case basis using sound scientific judgement. Examples are:

- A particle size determination does not need to be specific for a particular analyte and thus specificity is not relevant.
- In a water determination it should be demonstrated that the test is specific for water in the presence of the drug sample.

Linearity

The section on linearity in the ICH guidelines[1] is as follows:

ICH

2. Linearity

The linearity of an analytical procedure is its ability (within a given range) to obtain test results which are directly proportional to the concentration (amount) of analyte in the sample.

A linear relationship should be evaluated across the range (see section 3) of the analytical procedure. It may be demonstrated directly on the drug substance (by dilution of a standard stock solution) and/or separate weighings of synthetic mixtures of the drug product components, using the proposed procedure. The latter aspect can be studied during investigation of the range.

Linearity should be evaluated by visual inspection of a plot of signals as a function of analyte concentration or content. If there is a linear relationship, test results should be evaluated by appropriate statistical methods, for example, by calculation of a regression line by the method of least squares. In some cases, to obtain linearity between assays and sample concentrations, the test data may need to be subjected to a mathematical transformation prior to the regression analysis. Data from the regression line itself may be helpful to provide mathematical estimates of the degree of linearity.

The correlation coefficient, y-intercept, slope of the regression line and residual sum of squares should be submitted. A plot of the data should be included.

In addition, an analysis of the deviation of the actual data points from the regression line may also be helpful for evaluating linearity.

Some analytical procedures, such as immunoassays, do not demonstrate linearity after any transformation. In this case, the analytical response should be described by an appropriate function of the concentration (amount) of an analyte in a sample.

For the establishment of linearity, a minimum of 5 concentrations is recommended. Other approaches should be justified.

Verification of the calibration model

If an analytical method is going to be used to generate quantitative results, then a method of calibration is required. The purpose of the validation characteristic referred to as 'linearity' is to verify the calibration model. The term 'linearity' is slightly misleading because it implies that the relationship between the test results and the concentration of the analyte should be linear. This is not always the case as is acknowledged in the penultimate paragraph of section 2 in the guidelines, **immunoassays** being quoted as an example. In the FDA guidelines[4] on bioanalytical method validation the term 'Calibration/Standard Curve' is used instead of linearity.

In reality a large number of pharmaceutical analytical methods are based on linear relationships and the terminology is appropriate in these situations. Therefore, the term 'linearity' is used in the remainder of this book, but it is assumed that the reader understands that a linear calibration may not always apply.

Single point calibration

The most common type of calibration model encountered in the analysis of drug substances and drug products is known as a single point calibration, where the standard is prepared at one concentration level only. In fact, this model defines a two point calibration line where one point equals zero and the other the standard concentration. A single point calibration line is shown in Figure 5. The unknown concentration of an analyte in a sample is calculated by entering the response obtained for the analyte when analysed by the method into the equation of the calibration line, $y = mx + c$. In order to force the line through zero, and thus use a single point calibration, the value of 'c' must be negligible.

To verify the single point model, it is necessary to demonstrate that the value of the 'c' (the intercept) is negligible. This is achieved by measuring the response due to a number of different concentration levels across the range of the method and calculating the intercept for the best fit straight line. Since the aim of the investigation is to show that the line goes through zero it is not appropriate to include zero as a point in the investigation unless it is a measured value. For example, the absence of a peak in a chromatographic analysis cannot be interpreted as a zero result. However, the measurement of the blank in a UV assay is a measured value and may be included in an investigation of linearity. This applies to all calibration models, not just single point.

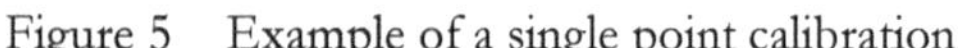

Figure 5 Example of a single point calibration

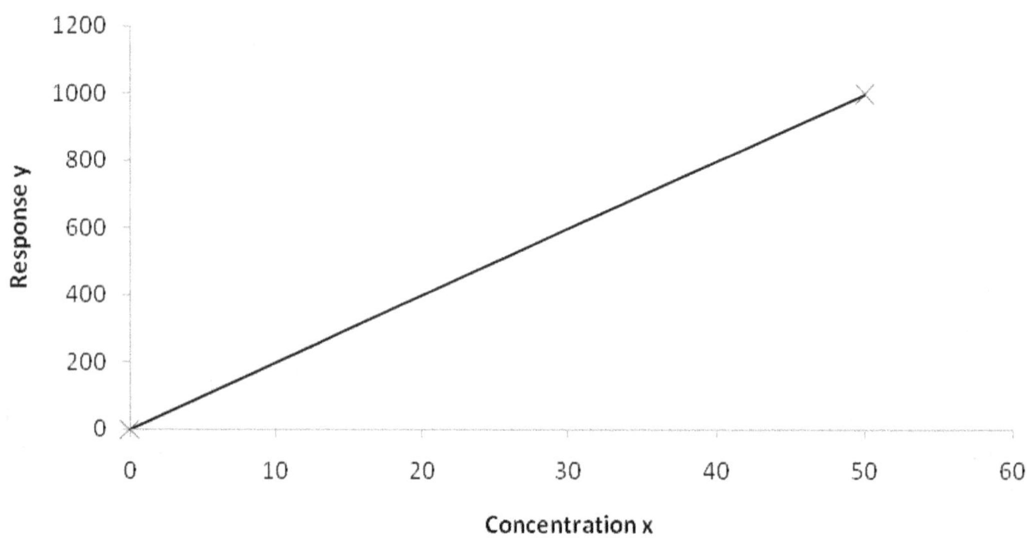

Multi-level calibration

For methods where a negligible intercept is difficult to achieve, for example bioanalytical methods which have varying matrices, a multi-level calibration is normally used. This involves measuring the response due to a number of standards at different concentration levels and constructing a calibration line each time the analysis is performed. To verify this model the response due to a number of different concentration levels (a greater number of levels than used routinely in the method) across the range of the method is measured and the relationship between the response and the concentration is investigated.

Regression analysis

The data obtained from analysis of the solutions prepared at a range of different concentration levels is usually investigated by plotting on a graph. If the method involves a mathematical transformation, e.g., log, square root or reciprocal, in order to obtain a linear calibration model then this should be performed prior to plotting the data. If method development and validation are being performed together and the calibration model is not yet defined the data should be plotted in order to investigate the relationship.

An example of data obtained for a linearity investigation is shown in Table 5 and is plotted in Figure 6.

Table 5 Data obtained from a linearity investigation

Concentration x	Response y		
	Measurement 1	Measurement 2	Measurement 3
10	215	216	214
20	380	376	378
30	631	634	637
40	850	842	844
50	996	999	990
60	1229	1241	1234
70	1423	1431	1417

Figure 6 Plot of the linearity data in Table 5

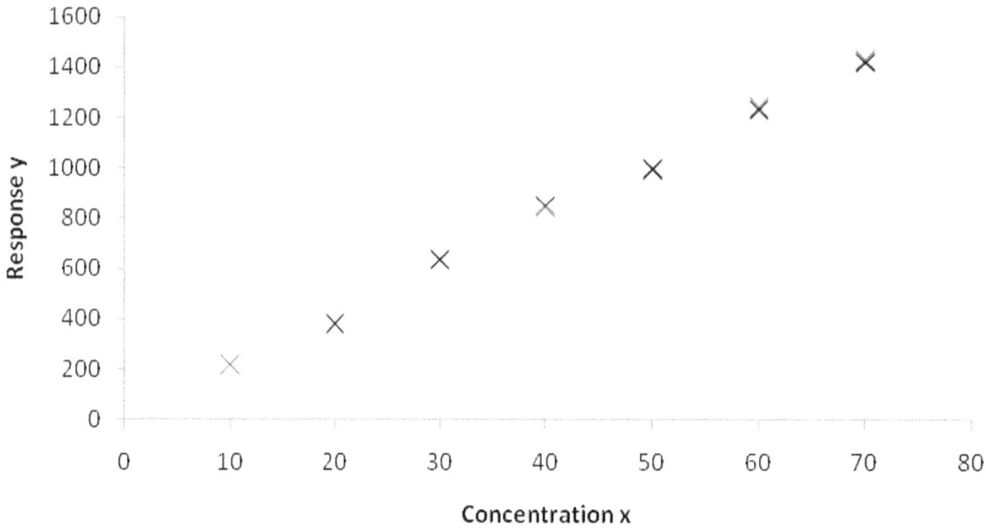

In this example the data clearly shows a linear relationship. To investigate further, the statistical technique of '**regression analysis**' is used, this refers to the fitting of equations to statistical data. The regression method of least squares produces the straight line which gives the best fit to the experimental data, obtained by minimising the distances between the experimental points and the line (these distances are called the residuals). Positive and negative residuals would cancel each other out so the squares of the residuals are used and these are minimised, hence 'least squares'. An important requirement of this type of regression is that the x-values are assumed to be error free, this is usually the case for the solutions that are prepared specifically to investigate linearity. When regression analysis is performed on the data in Table 5, the straight line which best fits the experimental data is obtained as shown in Figure 7.

Figure 7 Regression analysis performed on the data in Table 5

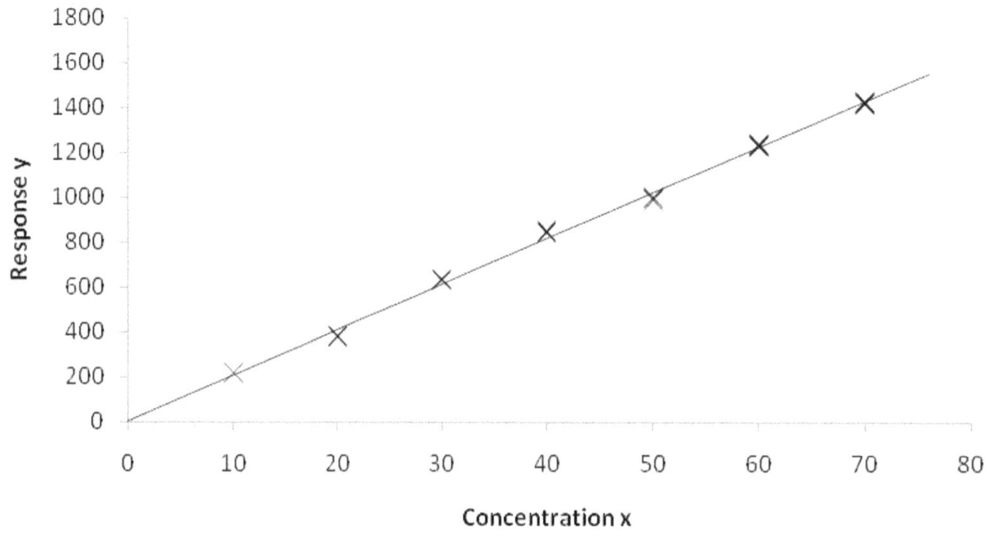

The output from regression analysis

Statistical data that is generated by regression analysis is summarised in Table 6.

Table 6 Summary of the statistics generated during regression analysis of the data in Table 5

Data	Description	Value calculated for the example
Equation of the line, $y = mx + c$	The relationship between the concentration, x and the response, y is described by the equation of the line.	$y = 20.36x + 3.62$
Intercept (c)	The value of y when x equals zero.	3.62
SE (intercept)	The standard error of the intercept, this value can be used to calculate the required confidence interval.	10.94
	95% Confidence interval =	-19.28 to 26.52
Slope (m)	The slope or gradient of the line relate to the steepness of the line and thus the relationship between concentration and response. It is equivalent to the response factor for the method, i.e., response divided by concentration.	20.36
SE (slope)	The standard error of the slope, this value can be used to calculate the required confidence interval.	0.24
	95% Confidence interval =	19.85 to 20.87
Coefficient of determination, r^2	The square of the correlation coefficient.	0.997
Correlation coefficient, r	The correlation between the predicted and observed values. This will have a value between 0 and 1, the closer the value is to 1, the better the correlation.	0.998
Regression SS	The regression sum of squares is the amount of variability in the response that is accounted for by the regression line.	3481479
Residual SS	The residual sum of squares is the variability about the regression line (the amount of uncertainty that remains). This is sometimes called the error sum of squares.	9550
Total SS	The total sum of squares is the total amount of variability in the response	3491029

It was previously stated that the value of the intercept must be negligible to justify the use of a single point calibration. The value for the intercept calculated from the linearity data in Table 6 is equal to 3.62. This may be demonstrated to be statistically negligible by means of the confidence interval, which is calculated for 95% confidence at -19.28 to 26.52. Since this interval includes zero, the true intercept can also be assumed to be zero and a single point calibration is justified for this method. An alternative approach is to express the intercept as a percentage of the analytical response at the target concentration, for example the 100% concentration level in the case of an assay. If this percentage is not significant then a single point calibration may be used. If the intercept value is not negligible then a multi-level calibration is normally used.

The coefficient of determination, r^2, is equivalent to the ratio of the regression SS and the total SS and thus is an expression of how much the variability in the response is fitted by the regression line. The correlation coefficient, r, is the square root of r^2. It is important to understand that the correlation coefficient is not a measure of linearity but rather a measure of how well the data fits the model. Therefore, a value of unity may be obtained for data which fits to a curve. However, in many pharmaceutical analyses the method is based on a known linear relationship (for example, Beer's law for methods which use UV as a detection technique) and in these cases the closer r is to unity then the better the data fits the linear model.

Residuals

An investigation into the deviation of the actual data points from the values predicted by the regression line provides useful information on the model. A visual evaluation of the pattern of the residuals can be performed by plotting the residuals against the corresponding x values. If the model is correct, the plot should show random behaviour without systematic pattern or regularities. The residual values from the linear regression performed previously are plotted in Figure 8. It can be seen that the data points are grouped by each set of replicate measurements but otherwise they are randomly distributed, and no trends are apparent.

Figure 8 Plot of residuals

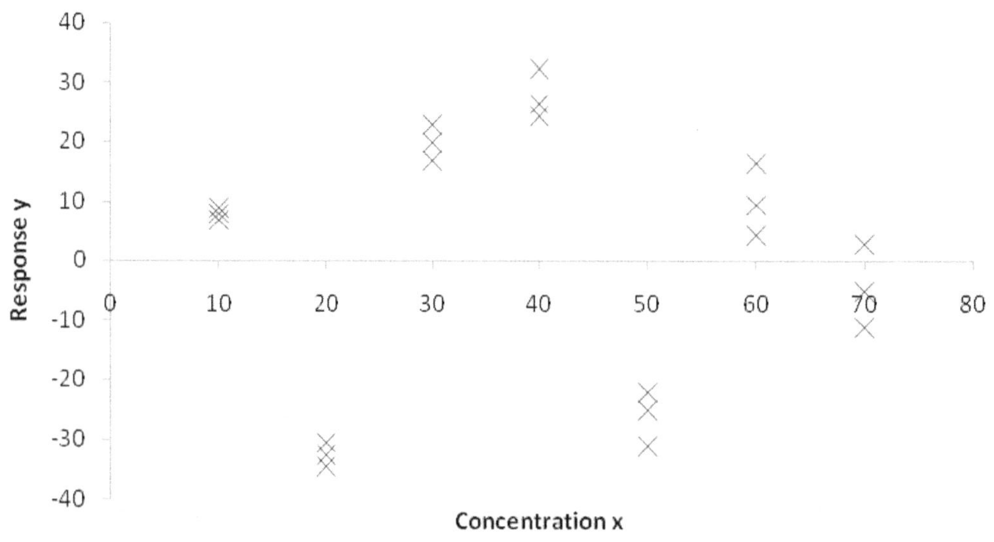

Weighting

The regression analysis described is based on a statistical model where the variation in the data at different concentration levels is the same. If the concentration range is more than two orders of magnitude, then it is likely that the variation in the data at different concentration levels will not be the same. This is known as **hetereoscedasticity**, or inhomogeneity of variances. In this situation simple regression using the method of least squares may not result in the best fit line. This is because larger concentrations with larger response values will require more reduction of their residuals and thus will have more impact on the minimisation of the sum of squares. Very small residuals may be neglected. To give all the data points equal importance weighted regression may be used where the data has a weighting factor applied. Common weighting factors used include the reciprocals of the actual variability (variance or standard deviation), or the respective concentration may be taken into account, i.e., $1/x$ or $1/x^2$.

Hetereoscedasticity may be encountered when using bioanalytical methods since a large concentration range may be required for analysis of drug and metabolites in biological fluids. These methods usually involve multi-level calibrations and thus regression may be used routinely when the method is used to derive the calibration line. If weighted regression is required for validation, then it is likely that it will need to be applied routinely. The FDA guidelines[4] on bioanalytical method validation state that 'the simplest model that adequately describes the concentration-response relationship should be used. Selection of weighting and use of a complex regression equation should be justified.'

For methods incorporating a large concentration range which use a single point calibration, weighted linear regression may be appropriate to demonstrate that the intercept value is negligible. A technical brief[8] by the Analytical Methods Committee (AMC) provides recommendations for the use of weighted regression. S. Toasaksiri et al.[9] describe the use of weighted linear regression on a capillary electrophoresis method used to separate non-steroidal anti-inflammatory drugs.

For more information on regression analysis the reader is referred to the chapter on fitting equations to data in the text 'Statistics for the Quality Control Chemistry Laboratory'[10] and also the text, 'Applied Linear Regression Models'.[11]

Non-linear calibration models

For methods which are not based on a linear relationship it may be possible to mathematically transform the data to linearity. For other methods which are inherently non-linear, such as microbiological and immunoassays, a non-linear regression model is required. The best model to use is considered on a case by case basis using scientific judgement to make the most appropriate selection. The reader is referred to the FDA guidelines[4] on bioanalytical method validation, the chapter on biological samples by K. A. Selinger in 'Development and Validation of Analytical Methods'[12] and textbooks on the topic of non-linear regression.[13,14]

Assay and impurity methods

For impurity methods linearity should be investigated over the appropriate range on solutions of the active pharmaceutical ingredient and also on specified impurities, i.e., impurities that are named on the specification for the drug substance or drug product. The linearity of the active pharmaceutical ingredient in this range is required since it is normally used to generate the calibration which is used to quantify unknown impurities. Specified impurities may also use this calibration. When investigating the linearity of specified impurities, the solutions prepared should contain: the impurity; the active pharmaceutical ingredient; and in the case of drug product, the excipients used in the formulation.

100% Standard methods

In section 3 of the ICH guidelines, it is stated that 'if assay and purity are performed together as one test and only a 100% standard is used, linearity should cover the range from the reporting level of the impurities to 120% of the assay specification.'[1] In this type of method the impurities (typical reporting limit of approximately 0.05%) are quantified using a standard at a much higher concentration level (100%). The standard used may be an external standard or may be the peak due to the active pharmaceutical ingredient in the sample being tested (known as area percent or normalisation).

To cover this full range from ~0.05% to 120% involves a linearity study over five orders of magnitude and it is likely that the data obtained will be hetereoscedastic. The options in this situation are to consider the whole range at the same time and possibly apply weighting to the data, or consider the impurity range and the assay range separately since data is expected only in these ranges in the normal use of the method. In the latter approach the slope obtained for the regression analysis on the impurity range is compared to that obtained for the assay range. The slope corresponds to the response factor for the analyte and should remain constant over the range of the method.

When using this approach, it is important to appreciate that the response factor depends on the response of the detection method for the analysis, if this is likely to vary from day to day and on different instruments then the analysis of the two ranges should be performed at the same time using the same analytical equipment. e.g., The response from a UV detector used for HPLC analysis may vary between instruments and may also depend on the age of the lamp in an individual instrument.

Bioanalytical methods

Multi-level calibration models are normally used for bioanalytical methods. A single point calibration model is usually unsuitable due to the biological matrix. The model may also involve weighting to account for hetereoscedasticity if the concentration range is over more than two orders of magnitude. The most common weighting function used for bioanalytical methods is $1/x$. A linear calibration model is preferred but a nonlinear model may be necessary for some methods.

Range

The section on range in the ICH guidelines[1] is as follows:

ICH

3. Range

The range of an analytical procedure is the interval between the upper and lower concentration (amounts) of analyte in the sample (including these concentrations) for which it has been demonstrated that the analytical procedure has a suitable level of precision, accuracy and linearity.

The specified range is normally derived from linearity studies and depends on the intended application of the procedure. It is established by confirming that the analytical procedure provides an acceptable degree of linearity, accuracy and precision when applied to samples containing amounts of analyte within or at the extremes of the specified range of the analytical procedure.

The following minimum specified ranges should be considered:

- for the assay of a drug substance or a finished (drug) product: normally from 80 to 120 percent of the test concentration;

- for content uniformity, covering a minimum of 70 to 130 percent of the test concentration, unless a wider more appropriate range, based on the nature of the dosage form (e.g., metered dose inhalers), is justified;

- for dissolution testing: +/-20 % over the specified range;

e.g., if the specifications for a controlled released product cover a region from 20%, after 1 hour, up to 90%, after 24 hours, the validated range would be 0-110% of the label claim.

- for the determination of an impurity: from the reporting level of an impurity* to 120% of the specification;

for impurities known to be unusually potent or to produce toxic or unexpected pharmacological effects, the detection/quantitation limit should be commensurate with the level at which the impurities must be controlled;

Note: for validation of impurity test procedures carried out during development, it may be necessary to consider the range around a suggested (probable) limit.

- if assay and purity are performed together as one test and only a 100% standard is used, linearity should cover the range from the reporting level of the impurities to 120% of the assay specification.

* see chapters "Reporting Impurity Content of Batches" of the corresponding ICH-Guidelines: "Impurities in New Drug Substances" and "Impurities in New Drug Products"

The working range of an analytical method is usually considered during the development of the method and will depend on the purpose of the method. The validated range is determined by the investigation of the validation characteristics of linearity, accuracy and precision. The ranges recommended for different types of methods are clearly defined in section 3 of the ICH guidelines. For other methods, the range can be derived by the same considerations, i.e., 20% above and/or below the specification limits as applicable.

The level at which an impurity is reported is defined in the ICH guidelines[15-17] relating to impurities in drug substances and drug products (reporting levels for individual residual solvents are also defined). The reporting level is dependent on the maximum daily dose of the drug. The requirements are summarised in Table 7. The reporting levels of impurities which are known to be unusually potent, or toxic are often controlled at a level much lower than those quoted in Table 7. The range of the method is therefore determined by the control level required. The levels in Table 7 do not apply to enantiomeric impurities due to the practical difficulties in quantification at those levels.

The upper value in the range for an impurity method is based on the specification for the impurity. It is likely that the specification limits for impurities will not be defined until a late stage in the development of a drug therefore probable specification limits for impurities are commonly used up to this point.

Table 7 Reporting thresholds for impurities in drug substances and drug products

	Maximum Daily Dose	**Reporting Threshold** (expressed as a percentage of the drug substance administered)
Drug Substance	≤2g/day	0.05%
	>2g/day	0.03%
Drug Product	≤1g/day	0.1%
	>1g/day	0.05%

Accuracy

The section on accuracy in the ICH guidelines[1] is as follows:

ICH

4. Accuracy

The accuracy of an analytical procedure expresses the closeness of agreement between the value which is accepted either as a conventional true value or an accepted reference value and the value found. This is sometimes termed trueness.

Accuracy should be established across the specified range of the analytical procedure.

4.1. Assay

4.1.1 Drug Substance

Several methods of determining accuracy are available:

a) application of an analytical procedure to an analyte of known purity (e.g., reference material);

b) comparison of the results of the proposed analytical procedure with those of a second well-characterized procedure, the accuracy of which is stated and/or defined (independent procedure, see 1.2.);

c) accuracy may be inferred once precision, linearity and specificity have been established.

4.1.2 Drug Product

Several methods for determining accuracy are available:

a) application of the analytical procedure to synthetic mixtures of the drug product components to which known quantities of the drug substance to be analysed have been added;

b) in cases where it is impossible to obtain samples of all drug product components, it may be acceptable either to add known quantities of the analyte to the drug product or to compare the results obtained from a second, well characterized procedure, the accuracy of which is stated and/or defined (independent procedure, see 1.2.);

c) accuracy may be inferred once precision, linearity and specificity have been established.

4.2. Impurities (Quantitation)

Accuracy should be assessed on samples (drug substance/drug product) spiked with known amounts of impurities.

In cases where it is impossible to obtain samples of certain impurities and/or degradation products, it is considered acceptable to compare results obtained by an independent procedure (see 1.2.). The response factor of the drug substance can be used.

It should be clear how the individual or total impurities are to be determined e.g., weight/weight or area percent, in all cases with respect to the major analyte.

4.3. Recommended Data

Accuracy should be assessed using a minimum of 9 determinations over a minimum of 3 concentration levels covering the specified range (e.g., 3 concentrations/3 replicates each of the total analytical procedure).

Accuracy should be reported as percent recovery by the assay of known added amount of analyte in the sample or as the difference between the mean and the accepted true value together with the confidence intervals.

In the definition of accuracy, it is stated that it 'is sometimes termed trueness'. In fact trueness refers to 'the closeness of agreement between the average value obtained from a large series of test results and an accepted value.'[18] It gives an indication of the bias for an analytical method, whereas accuracy relates to the absence of error in an individual result and is defined as 'the closeness of agreement between a test result and the accepted reference value'.[18] Accuracy is an expression of both trueness and precision since both these influence the result. The consequence of this distinction in definitions means that accuracy is determined for each of the individual samples prepared and analysed in the investigation but when the overall result is expressed as a mean, this value is actually the trueness of the method.

All accuracy investigations should be performed following the method which is being validated, this ensures that the results that are calculated are based on the calibration model in the method. This is important because the purpose of validation is to demonstrate the suitability of the routine analytical method.

Assay

Drug Substance

Section 4.1.1 provides three options for the demonstration of accuracy for a drug substance assay method.

The first option requires a reference standard of known purity which can be analysed using the method. In effect this will probably result in the same reference standard material being used to prepare the test solutions for the samples used to determine accuracy and the calibration standards. Recovery information obtained from these simple solutions is of questionable value.

The second option is to use a second well-characterised method. If such a method is available, then the results obtained by analysing the same sample using each method may be compared.

The third option, to infer from other validation characteristics, should be considered a last resort since it does not provide absolute measures.

Drug Product

Section 4.1.2 provides three options for the demonstration of accuracy for a drug product assay method.

The first option involves preparing a synthetic mixture of the drug product components which contains a known amount of drug substance, sometimes referred to as a 'pseudosample'. This sample is analysed using the method and the result may be expressed as a percentage of the known value, giving a percentage recovery, or linear regression may be performed on the data for actual amount versus estimated amount (an unbiased analysis has a slope of 1 and an intercept of zero). An example of an accuracy determination expressed as percentage recovery is shown in Table 8 and expressed as a linear regression in Figure 9. Note for the regression analysis that the slope is equal to 1.008 and the intercept is equal to -0.008, indicating an unbiased analysis.

Table 8 Example of percentage recovery results determined for an assay validation

Sample no.	Actual amount (mg/ml)	Amount found (mg/ml)	Recovery (%)
1	0.804	0.801	99.6
2	0.795	0.797	100.3
3	0.802	0.798	99.5
4	1.001	0.999	99.8
5	1.001	0.995	99.4
6	0.999	1.001	100.2
7	1.202	1.198	99.7
8	1.199	1.204	100.4
9	1.201	1.205	100.3
		Mean Recovery (%)	99.0
		95% Confidence interval	99.7 to 100.2
		RSD (%)	0.39

Figure 9 Linear regression performed on data in Table 8

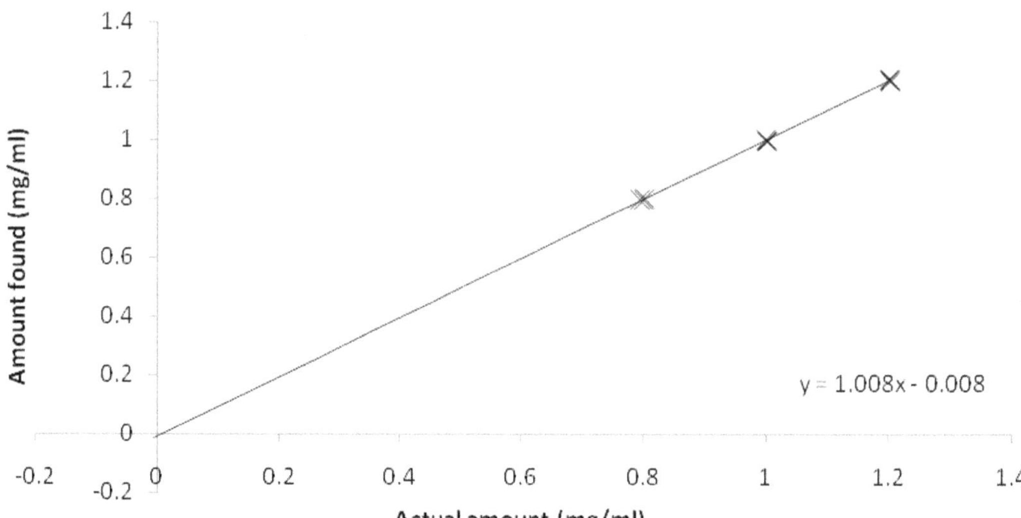

The dosage form of the drug product will determine how authentic the sample preparation for the accuracy samples can be. For many dosage forms, it is difficult to manufacture a single dosage unit which contains a known amount of active pharmaceutical ingredient in order to evaluate recovery. For example, if a solution of excipients is spiked with a solution of the active pharmaceutical ingredient to obtain the accuracy sample, then any sample preparation steps such as grinding, extracting, etc., are not included in the recovery information obtained. Possible implications of the differences between the sample preparation as directed in the method compared to the sample preparation for the accuracy samples should be considered.

The second option requires either a well-characterised second method, which may be difficult to obtain, or the addition of analyte to the drug product, which limits the range of the investigation to above the nominal content.

As is the case for drug substance, the third option, to infer from other validation characteristics, should be considered a last resort since it does not provide absolute measures.

Impurities
As for linearity, accuracy determination is only necessary for specified impurities. If samples of the impurity of interest are available, then accuracy may be evaluated by spiking samples of drug substance or drug product and calculating the recovery. The sample which is spiked should be free from the impurity of interest but if this is not possible then the amount present in the un-spiked sample should be quantified using the method and included in the calculations. Note that in section 7, acceptable accuracy is required at the 'minimum level at which the analyte can be quantified' and thus an accuracy investigation for an impurity method must include determination at the quantitation limit.

In situations where it is impossible to obtain samples of impurities, accuracy may be demonstrated using an independent analytical method. Additionally, an accuracy determination may be performed on the active pharmaceutical ingredient at the impurity level, prepared in the sample matrix. The purpose of this is to give confidence that impurities can be recovered at this low level in the presence of the sample components, with the obvious exception of the active pharmaceutical ingredient.

When performing accuracy on drug product possible implications of the differences between the sample preparation as directed in the method compared to the sample preparation for the accuracy samples should be considered (as in the case of the assay). For example, spiking with a solution of an impurity to obtain samples for accuracy does not demonstrate that the impurity can be extracted from the sample matrix.

Impurities are often quantified using a calibration model based on the active pharmaceutical ingredient. In the case of unknown or unavailable impurities the response factor of the active pharmaceutical ingredient is used to quantify the impurities. If an impurity is available, then the response factor relative to that of the active pharmaceutical ingredient can be determined and a correction included in the results calculations if appropriate.

Other methods

Sound scientific judgement should be used to evaluate the recovery requirements for a particular method and consideration given to the best way to demonstrate the recovery. For example, in the case of water determination by Karl Fischer analysis the sample may be susceptible to picking up water during handling and thus the best way to check accuracy may be to spike the titration vessel after the sample has been added. However, this does not demonstrate that the water can be extracted from, for example, a solid dosage form.

Recovery is of particular importance to analytical methods which are used to analyse cleaning samples. Cleaning samples which are obtained from swabbing cleaned manufacturing equipment are analysed to ensure that any residue is at an acceptable level. An important component of these methods is therefore the swabbing technique. Recovery should be determined during the development of the method. During the analytical method validation accuracy testing will include the demonstration of the recovery achieved by the swabbing technique.

Precision

The section on precision in the ICH guidelines[1] is as follows:

ICH

5. Precision

The precision of an analytical procedure expresses the closeness of agreement (degree of scatter) between a series of measurements obtained from multiple sampling of the same homogeneous sample under the prescribed conditions. Precision may be considered at three levels: repeatability, intermediate precision and reproducibility.

Precision should be investigated using homogeneous, authentic samples. However, if it is not possible to obtain a homogeneous sample it may be investigated using artificially prepared samples or a sample solution.

The precision of an analytical procedure is usually expressed as the variance, standard deviation or coefficient of variation of a series of measurements.

Repeatability

Repeatability expresses the precision under the same operating conditions over a short interval of time. Repeatability is also termed intra-assay precision.

Intermediate precision

Intermediate precision expresses within-laboratories variations: different days, different analysts, different equipment, etc.

Reproducibility

Reproducibility expresses the precision between laboratories (collaborative studies, usually applied to standardization of methodology).

Validation of tests for assay and for quantitative determination of impurities includes an investigation of precision.

5.1. Repeatability

Repeatability should be assessed using:

a) a minimum of 9 determinations covering the specified range for the procedure (e.g., 3 concentrations/3 replicates each);

or

b) a minimum of 6 determinations at 100% of the test concentration.

5.2. Intermediate Precision

The extent to which intermediate precision should be established depends on the circumstances under which the procedure is intended to be used. The applicant should establish the effects of random events on the precision of the analytical procedure. Typical variations to be studied include days, analysts, equipment, etc. It is not considered necessary to study these effects individually. The use of an experimental design (matrix) is encouraged.

5.3. Reproducibility

Reproducibility is assessed by means of an inter-laboratory trial. Reproducibility should be considered in case of the standardization of an analytical procedure, for instance, for inclusion of procedures in pharmacopoeias. These data are not part of the marketing authorization dossier.

5.4. Recommended Data

The standard deviation, relative standard deviation (coefficient of variation) and confidence interval should be reported for each type of precision investigated.

Precision levels

Precision is determined for a number of different levels during validation. These are:

1. System Precision
2. Repeatability
3. Intermediate Precision
4. Reproducibility

The total variability present at each of these precision levels increases from 1 to 4. This is a cumulative effect. For example, the variability at level 3, intermediate precision, consists of the variability due to the system precision plus that due to repeatability and also due to intermediate precision. A discussion of each precision level follows.

System precision

The first level is the system precision which investigates the variability of the measurement itself. This is also known as instrument/injection precision or injection repeatability. System precision is a better description of the variability since it is not all due to the injector. Although system precision is not referred to in the ICH guidelines it is an important component of system suitability testing and provides valuable data on the variability of the analytical system. An example of typical data obtained for the system precision of a HPLC analysis is shown in Table 9.

Measurement of precision

The precision results for the data in Table 9 are quoted as suggested in the ICH guidelines, i.e., standard deviation, relative standard deviation and confidence interval. It can be seen that standard deviation and the confidence interval are difficult to interpret without the mean and even then it is difficult to determine if they are acceptable precision values. The relative standard deviation is the ratio of the standard deviation to the mean, usually expressed as a percentage, and thus provides a standardised value which is meaningful for any dataset.

Table 9 An example of a system precision determination for a HPLC analysis

Injection number	Area response for analyte peak in standard
1	451677
2	450767
3	447653
4	452556
5	449336
6	448762
Mean	450125
Standard deviation	1861.3
%RSD	0.41
95% Confidence interval	448636 to 451615

Repeatability

The second precision level is repeatability, this provides a measure of the variability in the results obtained when using the analytical method to analyse samples taken from a homogenous batch in a single determination. In addition to the system precision, factors contributing to the variability include contributions from the sample preparation, such as weighing, pipetting, dilution, extraction etc. The analytical method must be applied fully to duplicate the way in which the method will be applied in routine use. Therefore, samples must be prepared individually rather than injecting the same solution multiple times (since this only evaluates system precision).

Where authentic samples are available then part (b) of section 5.1 is most appropriate. However, if artificially prepared samples are required then part (a) or part (b) may be followed. In this situation part (a) may be combined with the study of other validation characteristics such as accuracy and linearity. To calculate the precision for the nine determinations performed when following part (a), the results need to be expressed in the same format. For example, if the results are expressed as concentrations, then there will be three different sets of data, but if they are expressed as percentage recoveries then there will be one set of data for the precision calculation.

Assay

An authentic, homogenous sample is preferred for precision since this represents how the method will be applied in routine use. For drug substances a typical, representative batch may be selected and sampled as required. For drug products the nature of the dosage form will determine how the assay sample is prepared. For example, solid dosage forms such as tablets are typically milled and combined to provide a homogenous sample for the assay test. In general, the assay preparation may be used for the repeatability test but if this is not possible then the ICH guidelines permit the use of an artificially prepared sample.

Impurities

Authentic samples for impurity methods require that a sample is available with the impurity present at a level suitable for the investigation. If this is not available, then spiking may be necessary to obtain a suitable sample. Unfortunately, it may be difficult to spike the impurity into the sample in such a way that the analytical method can be applied in the same way as in routine use. The use of a spiking solution does not allow sample preparation steps of milling, extraction, etc to be evaluated. However, obtaining a homogenous sample by spiking a very small amount of solid into a solid matrix is extremely difficult. Stress studies may be useful to obtain samples which have degradation products present at suitable levels. It may be necessary to use more than one sample to assess repeatability when there is more than one impurity of interest.

The variability which may be expected in an impurity method is greater than that which would be expected in an assay method. This effect of increasing variability with decreasing concentration was observed by William Horwitz[19,20] when he assessed the results of thousands of collaborative trials (inter-laboratory method performance studies, mostly in food analysis). The relationship is expressed using the 'Horwitz Trumpet', shown in Figure 10. The lines are best regarded as confidence intervals. A remarkable aspect of this relationship is that the results seemed to obey this law regardless of the nature of the analyte and the test material, or the physical principle underlying the measurement method.

Figure 10 The 'Horwitz Trumpet'

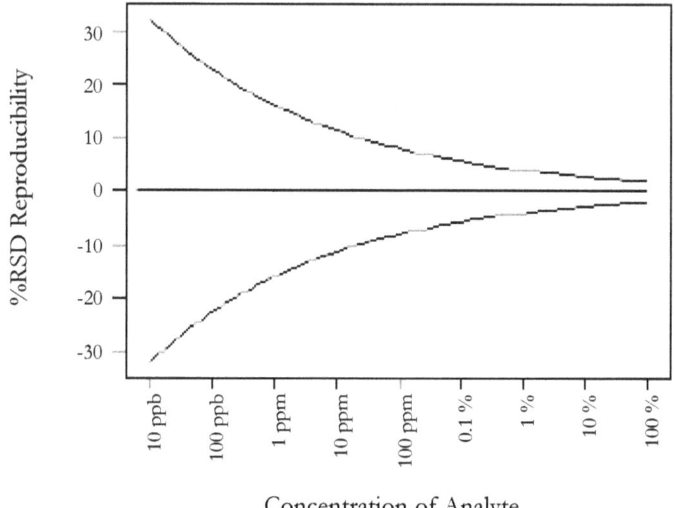

In methods used for pharmaceutical analysis the reason for increasing variability as the concentration decreases may be due to detector variability. When using chromatographic methods (commonly used for impurities determination) it is observed that the variability observed for results approaching the limit of quantification method increases rapidly. This is probably due to the integration of the peak. As the magnitude of the peak decreases the integration error increases.

Other methods

When assessing repeatability for other methods the preferred approach is to apply the analytical method to authentic samples. For some methods, such as content uniformity and dissolution, it is not usually possible to obtain a homogenous sample for repeatability. A possible approach for a solid dosage form in this situation is to weigh each individual tablet or capsule and use this value to correct the results (this assumes a homogenous distribution in the blend).

Intermediate precision

The third precision level is intermediate precision, this provides a measure of the variability in the results obtained when using the analytical method for routine analysis in the laboratory. In addition to the system precision and repeatability, factors contributing to the variability include contributions from the analyst operating the method, the time the analysis is performed, and the instrument used for the analysis. The investigation is based on the principles described under repeatability, but the analysis is performed a number of times altering the conditions under which it is performed. Historically, intermediate precision has also been referred to as 'ruggedness'.

The factors provided by the ICH guidelines as examples are days, analysts and equipment and many intermediate precision studies consider these three conditions only. However, it is stated in the guidelines that 'the extent to which intermediate precision should be established depends on the circumstances under which the procedure is intended to be used' and therefore the use of a method in a particular laboratory should be considered. Other factors which may be studied during intermediate precision include reagents, and for chromatographic techniques, columns.

Experimental design is encouraged in the ICH guidelines. This approach enables the evaluation of the effect of more than one factor at a time. In the example in Table 10 a fraction factorial design allows 4 runs to be performed in 2 days. Many different types of experimental design are available and may be used as long as the approach is based on sound scientific judgement and is justified.

As for repeatability, during intermediate precision the analytical method must be applied fully to duplicate the way in which the method will be applied in routine use. Therefore, each run must be performed independently.

Table 10 Example of experimental design of four runs for intermediate precision

Run 1	Run 2	Run 3	Run 4
Day 1	Day 1	Day 2	Day 2
Operator 1	Operator 2	Operator 1	Operator 2
Equipment 2	Equipment 1	Equipment 1	Equipment 2

Reproducibility

The fourth precision level is reproducibility, this provides a measure of the variability in the results obtained when using the analytical method for routine analysis in more than one laboratory. Factors contributing to the variability are similar to those for intermediate precision but with the added effects of a different location. Typically a reproducibility study is similar to that described for intermediate precision but conducted is in different laboratories.

Analytical methods are often transferred during the drug development process. Studies such as stability and clearance testing for clinical trials are often outsourced to contract research organisations and this involves ensuring that the contract research organisation can operate all the required methods to an acceptable standard. When the drug moves into the manufacturing phase the methods will need to be transferred from the development function to operations.

Although not required for marketing authorisation submissions, reproducibility may be subject to inspection during a regulatory authority audit and thus is an important validation characteristic. Useful references on the transfer of analytical methods are provided at the end of this chapter.[21,22]

An example of the results obtained for a precision study which utilised the experimental design proposed in Table 10 is shown in Table 11. The precision is expressed using the relative standard deviation of the data. Statistical tests which evaluate significance (difference) or equivalence may be used to compare the data obtained for the intermediate precision and reproducibility studies.

Table 11 Example of results obtained for a precision study

	Laboratory 1				Laboratory 2			
	Analyst 1		Analyst 2		Analyst 3		Analyst 4	
Sample	**Day 1 Equip 2**	**Day 2 Equip 1**	**Day 1 Equip 1**	**Day 2 Equip 2**	**Day 3 Equip 4**	**Day 4 Equip 3**	**Day 3 Equip 3**	**Day 4 Equip 4**
1	99.94	99.87	100.34	100.98	99.65	100.23	100.34	99.34
2	99.06	99.76	100.45	100.45	99.04	100.32	101.12	99.76
3	99.56	100.14	100.01	100.34	100.54	99.86	99.15	99.65
4	100.76	100.45	99.87	100.01	100.23	99.65	99.35	99.29
5	100.23	100.45	99.91	100.05	99.47	98.78	99.34	99.98
6	99.56	99.02	100.98	99.99	100.39	99.99	100.87	100.13
Mean	99.85	99.95	100.26	100.30	99.89	99.81	100.03	99.69
%RSD	0.60	0.54	0.42	0.38	0.59	0.56	0.86	0.34
Intermediate precision (n=24)								
Mean			**Laboratory 1**	100.09			**Laboratory 2**	99.85
%RSD				0.50				0.59
Reproducibility (n=48)								
Mean								99.97
%RSD								0.55

Accuracy, trueness and precision

Having considered accuracy, trueness and precision in the previous sections, a comparison of the meaning of these terms can now be made. Figure 11 shows a graphical representation[23] of their relationship to each other. It can be seen that an increase in the accuracy of the results is due to improving both precision and trueness and the effect is an overall decrease in uncertainty.

Figure 11 Accuracy, trueness and precision

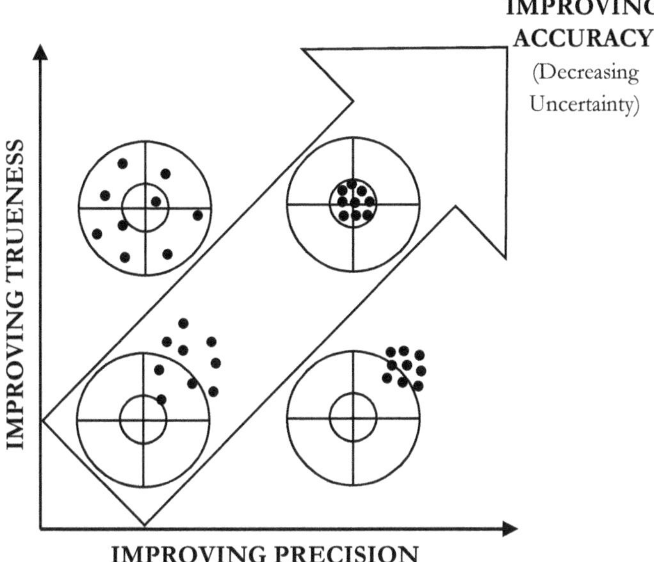

Detection Limit

The section on detection limit in the ICH guidelines[1] is as follows:

ICH

6. Detection Limit

The detection limit of an individual analytical procedure is the lowest amount of analyte in a sample which can be detected but not necessarily quantitated as an exact value.

Several approaches for determining the detection limit are possible, depending on whether the procedure is a non-instrumental or instrumental. Approaches other than those listed below may be acceptable.

6.1. Based on Visual Evaluation

Visual evaluation may be used for non-instrumental methods but may also be used with instrumental methods.

The detection limit is determined by the analysis of samples with known concentrations of analyte and by establishing the minimum level at which the analyte can be reliably detected.

6.2. Based on Signal-to-Noise

This approach can only be applied to analytical procedures which exhibit baseline noise.

Determination of the signal-to-noise ratio is performed by comparing measured signals from samples with known low concentrations of analyte with those of blank samples and establishing the minimum concentration at which the analyte can be reliably detected. A signal-to-noise ratio between 3 or 2:1 is generally considered acceptable for estimating the detection limit.

6.3 Based on the Standard Deviation of the Response and the Slope

The detection limit (DL) may be expressed as:

$$DL = \frac{3.3\sigma}{S}$$

where σ = the standard deviation of the response

S = the slope of the calibration curve

The slope S may be estimated from the calibration curve of the analyte. The estimate of σ may be carried out in a variety of ways, for example:

6.3.1 Based on the Standard Deviation of the Blank

Measurement of the magnitude of analytical background response is performed by analyzing an appropriate number of blank samples and calculating the standard deviation of these responses.

6.3.2 Based on the Calibration Curve

A specific calibration curve should be studied using samples containing an analyte in the range of DL. The residual standard deviation of a regression line or the standard deviation of y-intercepts of regression lines may be used as the standard deviation.

6.4 Recommended Data

The detection limit and the method used for determining the detection limit should be presented. If DL is determined based on visual evaluation or based on signal to noise ratio, the presentation of the relevant chromatograms is considered acceptable for justification.

In cases where an estimated value for the detection limit is obtained by calculation or extrapolation, this estimate may subsequently be validated by the independent analysis of a suitable number of samples known to be near or prepared at the detection limit.

The actual detection limit (sometimes known as the limit of detection, or LOD) for an analytical method may vary each time the analysis is performed. This is due to slight differences in the analytical equipment such as the intensity of the lamp in UV spectroscopy. Therefore, the detection limit determined from validation studies should be a reliable value which allows for the typical variation that may be experienced when operating the method.

When an impurities method is developed initially the detection limit which is required will be known since this is based on the levels defined in the ICH guidelines[15-17] relating to impurities in drug substances and drug products. The most common approach in this situation is to confirm the value experimentally and ensure it conforms to the signal to noise ratio requirements outlined in section 6.2 (assuming that the analytical technique exhibits baseline noise).

For methods which do not exhibit baseline noise, or where the detection limit needs to be determined, section 6.3 details an approach which uses the standard deviation of the response and the slope. The calculation is based on the calibration curve of the analyte. The value of x (detection limit) where y (the response) is due to the presence of analyte is found by multiplying the standard deviation of the blank, the regression line or the y-intercept by 3.3.

In 6.3.1 blank samples are analysed and the standard deviation of the results is determined. This approach is obviously only suitable where a response is obtained for a blank sample. For example, in the case of chromatographic methods, it would be expected that no peak is present in the blank (at the retention time of the analyte of interest) and therefore to obtain a measured value for the blank would necessitate the integration of noise, which would not result in meaningful results.

In 6.3.2 a calibration curve is constructed using samples in the range of the detection limit. Then the detection limit is determined using the standard deviation of either the regression line or the intercept. This approach is based on an unweighted linear regression and therefore the data must not be hetereoscedastic, as this would result in domination of the high concentrations with large responses and an incorrect, large detection limit would be calculated. A larger number of data increases the reliability of the calculated result and thus a minimum of eight concentrations is recommended for this study.

Quantitation Limit

The section on quantitation limit in the ICH guidelines[1] is as follows:

ICH

7. Quantitation Limit

The quantitation limit of an individual analytical procedure is the lowest amount of analyte in a sample which can be quantitatively determined with suitable precision and accuracy. The quantitation limit is a parameter of quantitative assays for low levels of compounds in sample matrices, and is used particularly for the determination of impurities and/or degradation products.

Several approaches for determining the quantitation limit are possible, depending on whether the procedure is a non-instrumental or instrumental. Approaches other than those listed below may be acceptable.

7.1. Based on Visual Evaluation

Visual evaluation may be used for non-instrumental methods but may also be used with instrumental methods.

The quantitation limit is generally determined by the analysis of samples with known concentrations of analyte and by establishing the minimum level at which the analyte can be quantified with acceptable accuracy and precision.

7.2. Based on Signal-to-Noise Approach

This approach can only be applied to analytical procedures that exhibit baseline noise.

Determination of the signal-to-noise ratio is performed by comparing measured signals from samples with known low concentrations of analyte with those of blank samples and by establishing the minimum concentration at which the analyte can be reliably quantified. A typical signal-to-noise ratio is 10:1.

7.3. Based on the Standard Deviation of the Response and the Slope

The quantitation limit (QL) may be expressed as:

$$QL = \frac{10\sigma}{S}$$

where σ = the standard deviation of the response

S = the slope of the calibration curve

The slope S may be estimated from the calibration curve of the analyte. The estimate of σ may be carried out in a variety of ways for example:

7.3.1 Based on Standard Deviation of the Blank

Measurement of the magnitude of analytical background response is performed by analyzing an appropriate number of blank samples and calculating the standard deviation of these responses.

7.3.2 Based on the Calibration Curve

A specific calibration curve should be studied using samples, containing an analyte in the range of QL. The residual standard deviation of a regression line or the standard deviation of y-intercepts of regression lines may be used as the standard deviation.

7.4 Recommended Data

The quantitation limit and the method used for determining the quantitation limit should be presented.

The limit should be subsequently validated by the analysis of a suitable number of samples known to be near or prepared at the quantitation limit.

The discussion for the detection limit applies equally to the quantitation limit with the following exceptions:

- The quantitation limit is the minimum concentration at which the analyte can be quantified with *acceptable accuracy and precision.* This means that the value which is calculated for the quantitation limit should be verified by preparing and analysing suitable samples.

- The multiplier for the standard deviation is equal to 10 for the calculation of the quantitation limit based on the standard deviation of the response and the slope.

Due to the variability of the actual detection limit and quantitation limit, samples of either or both may be prepared and included in the analysis each time the method is performed to ensure that the limits are valid.

The quantitation limit for bioanalytical methods[4] is usually referred to as the lower limit of quantification (LLOQ). It is required that the response due to the analyte at this level is at least five times that of the response due to the blank. The range of a bioanalytical method is usually defined by the concentration level at the LLOQ and at the upper limit of quantification (ULOQ).

Robustness

The section on robustness in the ICH guidelines[1] is as follows:

ICH

8. Robustness

The robustness of an analytical procedure is a measure of its capacity to remain unaffected by small, but deliberate variations in method parameters and provides an indication of its reliability during normal usage.

The evaluation of robustness should be considered during the development phase and depends on the type of procedure under study. It should show the reliability of an analysis with respect to deliberate variations in method parameters.

If measurements are susceptible to variations in analytical conditions, the analytical conditions should be suitably controlled or a precautionary statement should be included in the procedure. One consequence of the evaluation of robustness should be that a series of system suitability parameters (e.g., resolution test) is established to ensure that the validity of the analytical procedure is maintained whenever used.

Examples of typical variations are:

- stability of analytical solutions;

- extraction time.

In the case of liquid chromatography, examples of typical variations are:

- influence of variations of pH in a mobile phase;

- influence of variations in mobile phase composition;

- different columns (different lots and/or suppliers);

- temperature;

- flow rate.

In the case of gas-chromatography, examples of typical variations are:

- different columns (different lots and/or suppliers);

- temperature;

- flow rate.

Robustness is not included in the tabular summary of required validation characteristics to be tested in the ICH guidelines, however it is expected that it is 'considered at an appropriate stage in the development of the analytical procedure'.[1] Robustness is included in the summary table provided in the draft FDA guidelines[24] (see Table 14). Evaluation of the robustness of an analytical method is usually performed in two phases. Initial robustness testing is part of the method development process. A more formal study is then performed during validation of the method.

The factors which are investigated in robustness studies are method related, they are parameters which are defined in the method, e.g., temperature. Altering the value of these factors is a deviation from the method. In contrast, the factors which are investigated in intermediate precision (sometimes referred to as ruggedness) are non-method related factors, e.g., the analyst performing the analysis. When investigating these factors the method is followed in full without any deviations. Robustness factors are sometimes referred to as internal factors and those relating to intermediate precision as external factors. When an analytical method has been shown to have acceptable intermediate precision and robustness through the validation study, confidence is gained that the method can be used successfully in routine analysis. Also included in robustness testing is the evaluation of the stability of test solutions which are used in the analysis.

Some examples of robustness factors from available literature[1,25,26] for a variety of analytical techniques are listed below:

High Performance liquid Chromatography (HPLC)

- Influence of variations of pH in a mobile phase
- Influence of variations in mobile phase composition
- Different columns (different lots and/or suppliers)
- Temperature
- Flow rate
- Buffer concentration (ionic strength)
- Additive concentration
- Gradient slope
- Initial mobile phase composition
- Final mobile phase composition
- Injection volume
- Sample preparation (pH of solutions, reagent concentrations, etc.)
- Equilibration time
- Column age

- Detection wavelength
- Detector time constant

Gas Chromatography (GC)

- Different columns (different lots and/or suppliers)
- Temperature (column, injection and detector)
- Flow rate
- Heating rate
- Split ratio
- Column head pressure
- Sample preparation (pH of solutions, reagent concentrations, etc.)

Supercritical Fluid Chromatography (SFC)

- Composition of mobile phase
- Volume of modifier fluid in mobile phase
- Mobile phase additives
- Flow rate
- Different columns (different lots and/or suppliers)
- Injection volume
- Temperature
- System pressure
- Detector properties
- Equilibration time

Capillary Electrophoresis (CE)

- Voltage
- Injection
- Buffer or electrolyte concentration (ionic strength)
- Buffer or electrolyte pH

- Buffer stability
- Organic solvent concentration
- Surfactant concentration
- Selectivity additive concentration
- Capillaries (manufacturer, batch)
- Capillary conditioning time
- Cooling (heat removal)
- Sample preparation (pH of solutions, reagent concentrations, etc.)
- Temperature
- Detection wavelength

Ion Chromatography (IC)

- pH
- Temperature
- Flow rate
- Different columns (different lots and/or suppliers)
- Sample preparation (pH of solutions, reagent concentrations, etc.)

Spectroscopy

- Time constant
- Solvent
- pH of test solution
- Temperature
- Wavelength accuracy
- Slit width
- Sample preparation (pH of solutions, reagent concentrations, etc.)

From the list it can be seen that robustness testing depends very much on the analytical technique and associated equipment. It is recommended to examine at least those parameters which are part of the operational qualification (OQ) of the equipment. Then the factors investigated can be extended to those specific to the analytical method under evaluation.

Robustness during method development

When an analytical method is developed initially the most suitable conditions are determined experimentally, sometimes in combination with modelling techniques. The process of deriving these conditions usually results in identification of the critical factors which need to be controlled to ensure that the method is robust. As an example to demonstrate this process, the development of a HPLC method is considered.

The first stage of the method involves preparing the sample in a suitable way for analysis. This will depend on the nature of the sample, which may range from a relatively straightforward drug substance to a complex drug product, such as a metered dose inhaler. Equally important is the solubility of the analyte, this will determine which solvents are suitable for the analysis. An appropriate extraction method may need to be developed. In the selection of suitable conditions information on the effects of differing conditions will be obtained. Examples of conditions are sample weight, shaking or dissolution time, sonication time, temperature of sample preparation, extraction volume, wash volume, centrifugation time, pH of the solution, composition of the reagents and type of reagents. In addition, experiments may be performed to investigate which combination of conditions gives the best result, these experiments generate robustness data.

The next stage of the method requires the chromatographic conditions for the analysis. Initially a suitable stationary phase and mobile phase system need to be selected, often a column scouting approach is used to identify the most appropriate system. Then other conditions such as temperature and detection wavelength (for UV detection) are optimised. Whether the optimisation is performed experimentally, by changing conditions and monitoring the effect, or using a computer model, the process will identify which parameters are most important to achieve the desired separation and the effects of different values for these parameters.

If computer modelling is applied to determine the best conditions, then the model may be used to investigate the robustness of the method[25]. Factors may be altered and the effect on the separation observed. Typically, this approach may be used for the factors of mobile phase composition, flow rate, column temperature, pH, buffer concentration and column attributes such as particle size, but it will depend on how the model was generated.

This description of the HPLC method development process shows how robustness data is being generated throughout method development. This will apply to the development of methods using many different analytical techniques although the computer models referred to are predominantly used for chromatography methods.

Robustness during the validation study

The formal robustness study performed during analytical method validation investigates the parameters which have been shown to be critical during method development. The aim of the study is to obtain data to show that these parameters are robust within a specified range and to identify appropriate system suitability testing which ensures the validity of the method whenever it is used.

A large number of factors exist which could be investigated in a robustness study for an analytical method, as can be seen from the list on the preceding pages. To investigate each of the parameters while keeping the others at the nominal value (one factor at a time approach) involves a large number of experiments, e.g. 6 factors at 2 levels requires 12 experiments. Also, interactive effects which occur when two factors are altered cannot be detected, although in practice these are usually negligible. Typically, robustness experiments are performed using experimental design where the aim is to obtain as much information as possible from a limited number of experiments.

Experimental design

Experimental designs used for robustness investigations include fractional factorials and Plackett-Burman designs[26]. These are also called screening designs because they allow screening of a large number of factors. In recent literature related to robustness studies for pharmaceutical analysis the Plackett-Burman design is the type most commonly applied. The majority of studies reported are performed on chromatographic methods[27].

The steps that can be distinguished in applying an experimental design for robustness test are defined by Vander Heyden et al.[27,28] and are summarised below:

1. Selection of factors and their levels
2. Selection of an experimental design
3. Selection of responses
4. Definition of the experimental protocol and execution of experiments
5. Estimation of factor effects
6. Graphical and/or statistical analysis of the effects
7. Drawing conclusions and, if necessary, taking precautions or measures

Selection of factors and their levels

The selection of factors should be based on those identified as critical during the method development process. This will include any which could potentially have an effect on the results. A sound understanding of the analytical method will assist the selection of the most appropriate factors. Rather than just selecting factors at random from the list provided previously, careful consideration of why each factor is being chosen will ensure that the results are meaningful.

Usually the factors are examined at two extreme levels, the interval between the levels represents the variability that might occur for the factor. The interval may be based on the uncertainty with which a level can be set and reset or often it is based on the knowledge and experience of the analyst. The broader the interval, the more likely it is to have a significant effect, but if it does not have an effect then the method is robust over a large interval for the factor. A smaller interval will be less likely to have a significant effect but the factor will need to be controlled more strictly in the method. There are no guidelines regarding systematic selection of factor levels therefore the most appropriate levels are chosen for a particular method. Some examples of factors and their levels are provided in Table 12.

Table 12 Examples of factors and levels selected for a robustness study on a HPLC method

Factor	-1	Nominal value	+1
Flow rate of the mobile phase (mL/min)	1.4	1.5	1.6
pH of the buffer	6.5	6.8	7.1
Column temperature (°C)	23	28	33
Column manufacturer/type	A	A	B
% organic modifier at beginning of gradient (%)	24	25	26
% organic modifier at end of gradient (%)	41	43	45
Buffer concentration (mM)	22.5	25.0	27.5
Detection wavelength (nm)	260	265	270

For the wavelength factor, the position of the nominal wavelength with respect to the spectrum needs to be considered. If the nominal wavelength is at λ_{max}, the response due to wavelengths above and below this wavelength will both be lower than that obtained at the nominal value and thus the interval will not represent extremes. In this situation a solution is to compare one extreme to the nominal.

For qualitative factors, such as column manufacturer in Table 12, the nominal level is preferably compared with another, rather than two different levels. In the example the nominal column A is compared to B. Factors relating to chromatographic columns provide a particular challenge during robustness testing. The main factors relating to columns are:

- Age of the column, columns deteriorate as the number of injections they receive increases and there is a point in their lifetime when they are no longer suitable to run a particular analytical method.

- Stationary phase batch to batch variability, although manufacturers strive to reduce variability it can be analyte dependent and thus show up as a problem for particular methods.

- Manufacturer, this may be refer to the same stationary phase packed by different manufacturers, or may be a different stationary phase which is thought to be equivalent.

It is difficult to decide if these are internal or external factors, they relate to the method but the age or batch is not specified in the method, so they are external. These factors are often included in intermediate precision investigations. Including these factors in a two level robustness screening design may not provide a suitable range of extremes.

For the factor relating to the age of the column, assessing a new column and one which has been used for n injections will evaluate if significant effects are observed after n injections. However, it will not provide information on the expected lifetime of the column. This author recommends a separate study monitoring the system suitability results obtained for a column routinely used for the analytical method to investigate the effect of column age.

If two columns of differing batches of stationary phase are not significantly different this does not mean that subsequent batches will be equivalent. Therefore, a two level design is not adequate to assess the effect of batch to batch variability. To use a different type of stationary phase in an analytical method would require an assessment of equivalence since it is well documented that columns of the same nominal packing, e.g., c_{18}, from different manufacturers do not give the same separations. Therefore, to introduce this type of factor in a robustness study may not be appropriate.

Selection of an experimental design

Two-level screening designs, such as Plackett-Burman and fractional factorial designs are most common, an example of a Plackett-Burman design for the study of up to 11 factors in 12 experiments is shown in Table 14.

For Plackett-Burman designs the number of experiments is always a multiple of 4. The number of factors which can be studied in these designs is n-1. Most commonly used designs have n=8 or n=12, enabling study of up to 7 or 11 factors. For the example in Table 13, 8 factors are required and therefore a 12 experiment design would be used. In this case 3 factors are not defined and these may be assigned as dummy or imaginary factors.

Table 13 Plackett-Burman design to examine up to 11 factors (A-K) in 12 experiments, -1 = low factor level, +1 = high factor level

Experiment	Factors										
	A	B	C	D	E	F	G	H	I	J	K
1	+1	+1	-1	+1	+1	+1	-1	-1	-1	+1	-1
2	-1	+1	+1	-1	+1	+1	+1	-1	-1	-1	+1
3	+1	-1	+1	+1	-1	+1	+1	+1	-1	-1	-1
4	-1	+1	-1	+1	+1	-1	+1	+1	+1	-1	-1
5	-1	-1	+1	-1	+1	+1	-1	+1	+1	+1	-1
6	-1	-1	-1	+1	-1	+1	+1	-1	+1	+1	+1
7	+1	-1	-1	-1	+1	-1	+1	+1	-1	+1	+1
8	+1	+1	-1	-1	-1	+1	-1	+1	+1	-1	+1
9	+1	+1	+1	-1	-1	-1	+1	-1	+1	+1	-1
10	-1	+1	+1	+1	-1	-1	-1	+1	-1	+1	+1
11	+1	-1	+1	+1	+1	-1	-1	-1	+1	-1	+1
12	-1	-1	-1	-1	-1	-1	-1	-1	-1	-1	-1

Selection of responses

The responses may be related to the results obtained for the analytical method such as concentrations and recoveries or they may be system suitability responses. Examples of system suitability responses for chromatographic techniques are relative retention time, resolution and tailing factor.

Definition of the experimental protocol and execution of experiments

The order of experiments in a screening study is usually randomised, however practical reasons sometimes results in the experiments being blocked or sorted by one or more factors. For example, the column changeover required for the example in Table 13. The reason for the random order is to minimise the effects, such as time or drift, of uncontrolled factors. 'A time effect represents changes in a response, as a function of time, and these changes are larger than the experimental error. A drift effect is a special case of a time effect occurring when the response changes continuously in one direction (increases or decreases) as a function of time'.[27] An example of a possible time effect is a variation in the room temperature of the testing laboratory over the days taken to perform the study. An example of a drift effect is the aging of a chromatographic column over the course of the study. Randomisation will not always solve the problem of these effects, however it is possible to build anti-drift designs.[29]

The experiments are performed altering the experimental conditions as required and analysing selected samples in the normal way. The selection of samples for analysis will depend on the responses which will be measured.

Estimation of factor effects

The effect of the factor on the response is the difference of the average responses observed when the factor was at the high and low level. This is calculated for each response and factor.

Graphical and/or statistical analysis of effects

The graphical interpretation of important effects is typically applied with a normal probability plot. The statistical interpretation is based on identifying statistically significant effects usually derived from a t-test or an F-test.[27] Further information regarding the interpretation of results obtained from experimental design can be found in the literature.[27,30]

Drawing conclusions and, if necessary, taking precautions or measures

A method is considered robust when no significant effects are found for the responses. If a significant effect is observed, then the method should be controlled as appropriate. The study should assist in the selection of suitable system suitability criteria for the method.

Solution stability evaluation

For analytical methods that require the preparation of test solutions, it is necessary to evaluate their stability. This allows the use of solutions over a suitable period of time. This is important for a number of reasons. Sample solutions may not be analysed for a period of time after preparation in the case of automated analyses with long run times. To enable these types of analyses, the stability of the solutions needs to be established. In the case of standard solutions, the certified reference material from which they are prepared may be of limited supply and thus extending the time for the use of a single preparation will be beneficial. It is advisable to investigate solution stability during method development.

A typical approach to establish solution stability is to analyse the solution following the method, and then store it under the required conditions. It is re-analysed after a specified period of time. The original results and results obtained for the stored samples are compared to evaluate the stability. Alternatively, freshly prepared samples may be used for comparison. The comparison may take the form of an absolute measure, i.e., the re-analysis is within specified limits, or statistical tests based on confidence intervals may be used. Typical storage conditions include refrigerated and, stored protected from light, over a period of 24 hours to several days. Freeze and thaw studies may be required to support samples in biological matrices for bioanalytical methods.[4]

System Suitability

The section on system suitability in the ICH guidelines[1] is as follows:

ICH

9. System Suitability

System suitability testing is an integral part of many analytical procedures. The tests are based on the concept that the equipment, electronics, analytical operations and samples to be analyzed constitute an integral system that can be evaluated as such. System suitability test parameters to be established for a particular procedure depend on the type of procedure being validated. See Pharmacopoeias for additional information.

Section 9 is included in the ICH validation guidelines because the results of a validation study should be incorporated into the system suitability test for the analytical method. A full discussion of system suitability is included in the pharmacopoeias.[31-33]

An example of the type of information which can be obtained from validation studies is a precision evaluation for a chromatographic method. Precision investigations provide information on expected system precision, and also the variability in retention time, relative retention time, resolution, etc. which may be expected when operating the method using different operators, equipment and over time. This is supported by the information obtained from robustness studies regarding robust intervals for the critical operating parameters of the method.

Summary

1. **Specificity**
   - For identification methods all potential samples which could be mistaken for the sample of interest are investigated.
   - For other methods all potential interferences are investigated.
   - Samples may be forcibly degraded to obtain potential degradation products.
   - Peak purity assessment may be performed for chromatographic techniques.

2. **Linearity**
   - The calibration model for the analytical method is verified during the evaluation of linearity.
   - Calibration may be single point, where a single standard concentration is used, or multi-level where a series of standard concentrations are used.
   - The statistical technique of regression analysis is used to evaluate the calibration model.
   - The majority of methods have a linear relationship but non-linear relationships may be modelled, e.g., in bioanalysis.
   - Weighting may be required if the range of the analytical methods covers over two orders of magnitude.

3. **Range**
   - The range evaluated should be appropriate for the intended use of the method.
   - An acceptable degree of linearity, accuracy and precision should be obtained over the range of the method.

4. **Accuracy**
   - Samples are prepared which are representative of authentic samples but contain known amounts of analyte to evaluate accuracy.
   - The recovery of the amount of analyte is calculated by comparing the amount estimated by the method to the known amount.

5. **Precision**
   - There are 4 types of precision typically evaluated during analytical method validation: System precision, repeatability, intermediate precision and reproducibility.
   - Precision is usually expressed in terms of the standard deviation calculated for the data.

6. **Detection limit**
   - Typically the required detection limit is already known and it is demonstrated during analytical method validation.
   - The variability of the instrumentation needs to be considered when determining the detection limit.

Summary

7. Quantitation limit
- Typically the required quantitation limit is already known and it is demonstrated during analytical method validation.
- Accuracy and precision are demonstrated at the quantitation limit.

8. Robustness
- Robustness data is obtained during method development.
- A formal investigation of robustness using an experimental design may be performed as part of analytical method validation, using the critical parameters identified during method development.
- Solution stability is assessed for solutions which are prepared during the analysis.

9. System suitability
- The system suitability requirements for the analytical method should include any relevant parameters identified during the validation study.

References

1. International Conference on Harmonisation (ICH) of Technical Requirements for Registration of Pharmaceuticals for Human Use, Topic Q2 (R1): **Validation of Analytical Procedures: Text and Methodology**, 2005, www.ich.org.

2. G. Christian (Ed.), **Analytical Chemistry**, 5th ed., p. 2, Wiley, New York, 1994.

3. J. Vessman, R. I. Stefan, J. F. Van Staden, K. Danzer, W. Lindner, D. T. Burns, A. Fajgelj, H. Müller, *Pure Appl. Chem.* **73**, 1381-1386, 2001, '**Selectivity in Analytical Chemistry**'.

4. Guidance for Industry: **Bioanalytical Method Validation**, US Food and Drug Administration, Center for Drugs and Biologics, Department of Health and Human Services, 2001, www.fda.gov.

5. International Conference on Harmonisation (ICH) of Technical Requirements for Registration of Pharmaceuticals for Human Use, Topic Q1A (R2): **Stability Testing of New Drug Substances and Products**, 2003, www.ich.org.

6. International Conference on Harmonisation (ICH) of Technical Requirements for Registration of Pharmaceuticals for Human Use, Topic Q1B: **Stability Testing: Photostability Testing of New Drug Substances and Products**, 1996, www.ich.org.

7. D. W. Reynolds, K. L. Facchine, J. F. Mullaney, K. M. Alsante, T. D. Hatajik, M. G. Motto, *Pharm. Technol.* **27** (1), 48-56, 2002, '**Available Guidance and Best Practices for Conducting Forced Degradation Studies**'.

8. Analytical Methods Committee, AMC Technical Brief, 2007, No. 27, '**Why are we weighting**' (available on RSC website, www.rsc.org).

9. S. Toasaksiri, D. L. Massart, Y. Vander Heyden, *Anal. Chim. Acta* **416**, 29-42, 2000, '**Study of method validation criteria in a capillary electrophoresis method for the separation of non-steroidal anti-inflammatory drugs**'.

10. E. Mullins, '**Statistics for the Quality Control Chemistry Laboratory**', RSC, 2003, p247-307.

11. J. Neter, M. Kutner, C. Nachtsheim, W. Wasserman, '**Applied Linear Regression Models**', 3rd Ed., Irwin, 1996.

12. K. A. Selinger, '**Biological Samples**' in 'Development and Validation of Analytical Methods', Eds. C. M. Riley, T. W. Rosanske, Elsevier Science Ltd., 1996, p249-292.

13. D. M. Bates, D. G. Watts, '**Nonlinear Regression Analysis and its Applications**', Wiley, 2007.

14. H. Motulsky, A. Christopoulos, '**Fitting Models to Biological Data Using Linear and Non-linear Regression: A Practical Guide to Curve Fitting**', Oxford University Press Inc., USA, 2004.

15. International Conference on Harmonisation (ICH) of Technical Requirements for Registration of Pharmaceuticals for Human Use, Topic Q3A(R2): **Impurities in New Drug Substances**, 2006, www.ich.org.

16. International Conference on Harmonisation (ICH) of Technical Requirements for Registration of Pharmaceuticals for Human Use, Topic Q3B(R2): **Impurities in New Drug Products**, 2006, www.ich.org.

17. International Conference on Harmonisation (ICH) of Technical Requirements for Registration of Pharmaceuticals for Human Use, Topic Q3C(R3): **Impurities: Guidelines for Residual Solvents**, 2002, www.ich.org.

18. ISO 5725, **Application of Statistics – Accuracy (Trueness and Precision) of the Results and Methods of Measurement – parts 1 to 6**, International Organisation of Standardization (ISO), Geneva, 2006.

19. W. Horwitz, L. R. Kamps, K. W. Boyer, *J. Assoc. Off. Anal. Chem.*, **63**, 1344 (1980).

20. Analytical Methods Committee, AMC Technical Brief, 2004, No. 17, '**The amazing Horwitz function**' (available on RSC website, www.rsc.org).

21. M. Broughton, J. Ermer, '**Transfer of Analytical Procedures**', in 'Method Validation in Pharmaceutical Analysis, a Guide to Best Practice', Eds J. Ermer, J. Miller, Wiley, 2005, p281-300.

22. J. Vial, A. Jardy, P. Anger, A. Brun, J. Menet, *J. Chromatogr.* **A815**, 173-182 (1998), '**Methodology for transfer of liquid chromatography methods based on statistical considerations**'.

23. Analytical Methods Committee, AMC Technical Brief, 2003, No. 13, '**Terminology – the key to understanding analytical science. Part 1: Accuracy, precision and uncertainty**' (available on RSC website, www.rsc.org).

24. Draft guidance for industry: **Analytical Procedures and Methods Validation**, US Food and Drug Administration, Center for Drugs and Biologics, Department of Health and Human Services, 2000, www.fda.gov.

25. G. Kleinschmidt, '**Robustness**', in 'Method Validation in Pharmaceutical Analysis, a Guide to Best Practice', Eds. J. Ermer, J. Miller, Wiley, 2005, p120-169.

26. B. Dejaegher, Y. V. Heyden, *LC•GC Eur.*, **20**(10), 526-532, 2007, '**Screening Designs (Part I) – Types and Properties**'.

27. B. Dejaegher, Y. V. Heyden, *J. Chromatogr.* **A1158**, 138-157, 2007, '**Ruggedness and robustness testing**'.

28. B. Dejaegher, Y. V. Heyden, *LC•GC Eur.*, **19**(7), 418-423, 2006, '**Robustness Tests**'.

29. J. L. Goupy, '**Methods for experimental design, principles and applications for physicists and chemists**', in 'Data handling in science and technology – volume 12', Eds. B. G. M. Vandeginste , S. C. Rutan, Elsevier, 1993, p159-177, 421-429.

30. Y. V. Heyden, D. L. Massart, '**Review of Robustness in Analytical chemistry**', in 'Robustness of analytical chemical methods and pharmaceutical technological products', Eds. Y. V. Heyden, D. L. Massart 'M. W. B. Hendriks, J. H. De Boer, A. K. Smilde, Elsevier, 1996, p79-147.

31. United States Pharmacopeia (USP) 30, Chromatography <621>.

32. European Pharmacopoeia (EP) 5, Chromatographic Separation Techniques, 2.2.46.

33. Japanese Pharmacopoeia (JP) 15, Liquid Chromatography, 27.

Further Reading

J. Ermer, J. Miller, '**Method Validation in Pharmaceutical Analysis, a Guide to Best Practice**', Wiley, 2005.

D. M. Bliesner, '**Validating Chromatographic Methods: A practical guide**', Wiley, 2006.

P. C. Meier, R. E. Zünd, '**Statistical Methods in Analytical Chemistry**', Wiley, 2000.

J. Miller, J. C. Miller, '**Statistics and Chemometrics for Analytical** Chemistry', 5th ed., Prentice Hall, 2005.

E. Mullins, '**Statistics for the Quality Control Chemistry Laboratory**', RSC, 2003.

M. W. B. Hendriks, J. H. De Boer, A. K. Smilde (editors), '**Robustness of analytical chemical methods and pharmaceutical technological products**', Elsevier, 1996.

C. C. Chan, Y. C. Lee, H. Lam, X. Zhang, '**Analytical Method Validation and Instrument Performance Verification**', Wiley, 2004.

C. M. Riley, T. W. Rosanske, eds., '**Development and Validation of Analytical Methods**', Pergamon, 1996.

Notes

The Validation Protocol

The planning and design of a validation study involves a number of steps. First, the appropriate validation characteristics, as described in Chapter 2, are selected for the analytical method being studied. Next, the experiments which are required for each characteristic are defined including the concentration levels and the number of replicates to be tested. Then the required results and **acceptance criteria** are defined for each test.

Contents of the Validation Protocol

The validation protocol details the design of the validation study. Its purpose is to provide information on which characteristics will be tested during the study, how the experiments will be performed, and what results will be calculated. Typical information in a validation protocol may include:

- Details regarding the analytical method (or methods) to be validated.
- The validation characteristics which will be investigated for the method(s).
- Details on how the experiments will be performed including: the type and number of solutions to be prepared; how the solutions should be prepared; how measurements should be performed.
- Details of the results which will be calculated (with directions if appropriate).
- The acceptance criteria which will be applied to the results.
- Details of the reference materials which will be used in the validation study.
- Details of batches of the material used during the validation study. In the case of precision studies, particular representative batches may be selected.
- Details of the equipment which will be used in the study.
- Details of responsibilities and required signatures.

Validation Characteristics to be Studied

A tabular summary which details the validation characteristics that should be applied for different types of methods is included in the ICH guidelines[1]. This table is extended in the draft FDA guidelines[2] on 'Analytical Procedures and Method Validation' to include specific tests and robustness, and is reproduced below, see Table 14.

Table 14 Recommended validation characteristics of the various types of tests[2]

Type of analytical procedure	Identification	Testing for impurities		Assay Dissolution (measurement only),	Specific Tests
Characteristic		Quantitative	Limit	Content/potency	
Accuracy	-	+	-	+	+[4]
Precision					
Repeatability	-	+	-	+	+[4]
Intermediate precision	-	+[1]	-	+[1]	+[4]
Specificity	+[2]	+	+	+[5]	+[4]
Detection Limit	-	-[3]	+	-	-
Quantitation Limit	-	+	-	-	-
Linearity	-	+	-	+	-
Range	-	+	-	+	-
Robustness	-	+	-[3]	+	+[4]

NOTE:

- \- Signifies that this characteristic is not normally evaluated.
- \+ Signifies that this characteristic is normally evaluated.

1 In cases where reproducibility has been performed, intermediate precision is not needed.

2 Lack of specificity for an analytical procedure may be compensated for by the addition of a second analytical procedure.

3 May be needed in some cases.

4 May not be needed in some cases.

5 Lack of specificity for an assay for release may be compensated for by impurities testing.

The 'Type of analytical procedure' in the table, i.e., identification, assay, testing for impurities and specific tests, was defined previously in Chapter 1. The table lists the validation characteristics which are regarded as most important for the different types of methods. The ICH guidelines[1] state that 'the list should be considered typical for the analytical procedures cited but occasional exceptions should be dealt with on a case-by-case basis'. Laboratories which perform analytical method validation typically have in-house guidelines or standard operating procedures to describe the approach which should be taken for validation studies. These documents are usually based on the ICH guidelines and provide additional information on the organisational policy regarding which validation characteristics should be investigated for particular types of analytical methods.

Experimental Procedure and Acceptance Criteria

When the relevant validation characteristics have been identified, the experimental procedure which will be used to investigate those characteristics needs to be defined. Minimum requirements for the number of determinations and replicates which should be used are provided in the ICH guidelines. Also required is the rationale for assessing the results which are generated from the study. The acceptance criteria selected may be based on an absolute value of practical relevance or a statistical test may be used. The experimental procedure for performing a validation study and the relevant acceptance criteria are discussed for each validation characteristic below. The values quoted are suggestions provided for the purposes of orientation. Therefore, they should be used with caution and may not be appropriate for a particular method.

Specificity

The aim of the specificity investigation is different for identification methods when compared to the aim for assay and impurity methods, due to the difference in the nature of the methods. For identification, specificity demonstrates that the method does not give a positive response for samples other than the one of interest. However, for assay and impurity methods, specificity demonstrates that the response due to the analyte of interest in the sample is not affected by potential interferences which may also be present in the sample. The result of this is that the approach used for the experimental part of the investigation and the acceptance criteria applied for each type of method differs.

Experimental

For identification methods samples will be selected which could be mistaken for the sample of interest. This selection requires knowledge of the analytical method, the manufacturing process for the material in question and knowledge regarding other processes using the same plant. The samples may include materials which are structurally similar or closely related to the analyte, e.g., isolated intermediates for the drug substance. Samples should be prepared in a similar way to the sample of interest and analysed as per method.

For assay and impurity methods the samples tested will contain materials which are potentially present during routine analysis and may interfere with the result, e.g., impurities, excipients, etc. Depending on the stage of development, impurities may not always be available in an isolated form. Therefore, samples investigated during specificity may be prepared from degraded material which has been subjected to stress testing or from material known to contain high levels of an impurity of interest. The level of the impurity in the sample tested should be typical of that expected in authentic samples since specificity may be concentration dependent. An example of this is in chromatographic analyses where a small impurity peak may be well resolved, but when the size of the peak increases the resolution decreases, and the separation is no longer adequate.

For bioanalytical methods the FDA guidelines[3] recommend that blank samples of the appropriate biological matrix should be obtained from at least six sources and each blank sample tested. Samples of other potentially interfering materials, e.g., metabolites, should also be tested.

Acceptance criteria

The acceptance criterion for an identification method is that the method can discriminate between the sample of interest and other samples with which it could be confused. For particular methods additional information could be supplied, such as comparison with a reference standard (e.g., IR spectroscopy) or an absolute value (e.g., melting point).

For assay, impurity and bioanalytical methods the acceptance criterion is that the method can discriminate between the analyte of interest and potential interferences which may be present in the sample. Additional information may relate to the extent of discrimination. In the case of a chromatographic method, separation of the peaks is the criteria for assessing specificity and thus the resolution of two peaks can be used as acceptance criteria, e.g., resolution must be greater than 1.5.

Linearity

Experimental

It is recommended in the ICH guidelines that a minimum of 5 concentrations are used for the determination of linearity. These may be prepared from 'the drug substance (by dilution of a standard stock solution) and/or separate weighings of synthetic mixtures of the drug product components'.[1] If a standard stock solution is used then it is important that the solubility of the compound is taken into account. The concentrations selected for the linearity investigation should cover the range of the method and are typically evenly spaced out over the range being evaluated. A point to note is that if the concentration range is over two orders of magnitude, then it may lead to weighted regression analysis being required.

When the method is applied to the linearity standards, single or replicate measurements for each concentration may be performed. If the system precision of the method is high, i.e., the variability of measurements is high, then replicates are advised so that the variability is taken into account in the regression analysis. However, if the precision is low, i.e., the variability of measurement is low, as in the case of many assay methods performed using HPLC, then single measurements (or injections in the HPLC example) should be adequate.

For assay methods based on a single point calibration model the concentration range investigated during linearity is typically over the range 80 to 120% of the nominal concentration. Since the value at the intercept is required to prove the validity of the model the range is actually 0 to 120% and this involves extrapolation over the relatively large range of 0 to 80%. To reduce any detrimental effects of the extrapolation (which may result in an erroneous non-zero intercept) it is a good idea to include an extra linearity standard at around 40%.

For bioanalytical methods, the FDA guidelines states that a minimum of 6 standards should be used to define the standard curve, in many cases 6 to 8 standards are adequate, but more standards may be required for non-linear relationships. The concentration of the standards should be selected to cover the required range including the LLOQ and are typically evenly spaced over the range.

Acceptance criteria

The overall acceptance criterion for linearity is that the validity of the calibration model has been demonstrated. The supporting information includes a residual plot which shows random scatter and no systematic trend and, for single point calibration models, the 95% confidence limits for the intercept should include zero or the intercept expressed as a percentage of the nominal concentration used in the method should not be greater than 2%.

The value of r is commonly used as an acceptance criterion, e.g., $r > 0.999$ for an assay method, but the value of r depends on the number of values and concentration range being evaluated therefore it is not suitable as a quantitative measure of linearity. However, it is very useful to assess how well the data fits to the regression line.

Range

Experimental

The range is defined by the experiments performed for linearity, accuracy and precision and thus has no specific experimental requirements.

Acceptance criteria

The acceptance criterion is that the analytical method 'provides an acceptable degree of linearity, accuracy and precision when applied to samples containing amounts of analyte within or at extremes of the specified range'.[1]

Accuracy

Experimental

It is recommended in the ICH guidelines that accuracy is 'assessed using a minimum of 9 determinations over a minimum of 3 concentration levels covering the specified range. (e.g., 3 concentrations/ 3 replicates each)'.[1] For assay methods a typical approach is to prepare 3 replicates at 80% of the nominal concentration, 3 at 100% and 3 at 120%, thus covering the range of the method. For impurity methods the required range is from the quantitation limit to 120% of the specification limit and so one approach would be to prepare 3 replicates at the quantitation limit, 3 at 50% of the specification limit and 3 at 120% of the specification limit. Alternatively, 9 different concentrations could be prepared. All accuracy samples are analysed using the method in the same way as it would be applied for routine analysis. If a second method is being used to investigate accuracy, then the same samples are analysed using this method.

For bioanalytical methods accuracy is assessed using a minimum of five determinations per concentration and a minimum of 3 concentrations in the range of expected concentrations is recommended.

Acceptance criteria

The acceptance criteria for the recovery of the accuracy samples are usually based on an acceptable range for the mean, e.g. for an assay method the criteria may be that the mean is within 98 - 102%, or for an impurities method, within 90 - 110%. Acceptable ranges may also be defined for the individual accuracy results. If linear regression is performed using the values of actual concentration versus the estimated amount, then acceptance criteria may be based on the slope and intercept, e.g., slope is within 0.98 to 1.02 and the 95% confidence interval includes 1 for an assay method and slope is within 0.9 to 1.1 and the 95% confidence interval includes 1 for an impurity method. When a second method is used for accuracy, the results calculated by the two methods are compared and the acceptance criteria may be based on an absolute range, as previously, or a statistical equivalence test may be used, such as a t-test.

For bioanalytical methods the deviation of the mean from the true value serves as the measure of accuracy. The mean value at each concentration should be within 15% of the actual value except at LLOQ, where it should not deviate by more than 20%.

Precision

Experimental

It is recommended in the ICH guidelines that repeatability is 'assessed using a) a minimum of 9 determinations covering the specified range for the procedure (e.g., 3 concentrations/ 3 replicates each)' or b) a minimum of 6 determinations at 100% of the test concentration.'[1] The use of authentic samples is preferred for precision studies and this usually means that the option b approach is taken. For assay methods the precision study is straightforward if a homogenous sample is available, 6 replicates of a representative batch are analysed. For impurity methods a batch of drug substance or drug product (as applicable) may need to be selected on the basis of the presence of the impurity of interest at a typical level, or a degraded sample may be used. If a synthetic sample is required then the accuracy and precision investigations can be performed together.

Although no recommendations are made in the ICH guidelines regarding the number of determinations to use for intermediate precision and reproducibility, the same approach as for repeatability is typically used. The variables to consider, e.g., days, instruments, etc., were discussed previously in Chapter 2. System precision is typically assessed using 6 measurements.

For bioanalytical methods precision is measured using a minimum of 5 determinations per concentration. A minimum of 3 concentrations in the range of expected concentrations is recommended.

Acceptance criteria

The most common acceptance criteria for precision are based on the % RSD for the data. Typical limits for system precision, repeatability and intermediate precision/ reproducibility for an assay method are < 1%, < 2% and < 3% respectively, and for impurities methods typical values are < 2%, < 10% and < 15% respectively. For bioanalytical methods, the precision expressed as % RSD for each concentration should not exceed 15%, except for the LLOQ, where it should not exceed 20%.

Detection Limit and Quantitation Limit

Experimental

The section on the validation of the detection limit and the quantitation limit in the ICH guidelines refers to analysis of 'a suitable number of samples' which are known to be near or prepared at the detection limit/quantitation limit. The suitable number will be dependent on the variability of the method at low concentrations, 6 samples provide a reasonable number to provide confidence in the detection limit and to give data on precision, and possibly accuracy, for the quantitation limit. Samples of available impurities should be spiked into drug substance or drug product to obtain representative samples, care must be taken to ensure that the impurity is not present in the drug substance or drug product being used. If the limit is being determined using the active pharmaceutical ingredient (if impurities are not available and to estimate the limits for unknown impurities) then the sample should be as representative as possible, e.g. drug product excipients should be present.

For the majority of impurity methods, the desired detection limit and quantitation limit will be already known, defined by the ICH requirements for impurities in drug substance and drug products, therefore the investigation will consist of preparing samples at the desired level and analysing them using the method to demonstrate that they are suitable. Since most impurities methods are based on chromatographic techniques the signal-to-noise approach is commonly used.

For other impurity and trace analysis methods the determination of the detection limit and quantitation limit may be achieved using the approaches discussed in Chapter 2. If the standard deviation of the blank is used then 6 samples would be suitable. If the approach using a calibration curve is utilised, then it is important that the concentration levels used for the determination are in the range of the expected detection limit and/or quantitation limit, the use of 8 concentration levels to generate the calibration curve are recommended. For bioanalytical methods the LLOQ should be established using at least five samples.

Acceptance criteria

The acceptance criteria for the detection limit or quantitation limit usually relates to their concentration level. For example, methods for organic impurities in drug substances require reporting thresholds of 0.03% or 0.05% nominal concentration, corresponding to a daily dose of >2g/day and ≤2g/day respectively (see Table 7). Thus, this is the required quantitation limit and the detection limit should be equal to quantitation limit divided by three. For other methods the level specified should be based on the control levels required.

The acceptance criteria for an impurities method, relating to the precision and accuracy required for the quantitation limit, is typically about 10 to 20% RSD for precision and about 90 to 110% recovery for accuracy. For bioanalytical methods the LLOQ should be at least 5 times the response compared to the blank response and the analyte peak should be identifiable, discrete, and reproducible, with a precision of 20% and accuracy of 80 to 120%.

Robustness

Experimental

The design of a robustness study follows the steps detailed in Chapter 2. The actual samples which should be analysed for each experiment will depend on the responses which are selected for the study (step 3). Typically, the analysis performed for each of the robustness experiments will be similar to that for repeatability and will include typical representative samples, which are calibrated using the model in the method, and also a system suitability solution if applicable.

Acceptance criteria

The overall acceptance criterion for robustness is that no significant effects are found for any of the factors investigated, or effects observed during the investigation are controlled by the system suitability requirements for the method.

Stability

Experimental

Samples prepared for repeatability may be used for test solution stability if a homogeneous sample is used. Storage conditions of room temperature and refrigerated are typical. The results are compared to the original analysis, or freshly prepared control samples may be used to compare against the stored samples at the time of reanalysis. The use of 3 samples is usually adequate. For bioanalytical methods the storage conditions are more extensive to account for the effects of the biological matrix. For freeze and thaw, and short and long term temperature studies 3 aliquots each of high and low concentration are assessed.

Acceptance criteria

The acceptance criteria for test solution stability may be based on a comparison of the stored samples against freshly prepared controls. Typical criteria are within 2% for an assay method and within 15% for a bioanalytical method. Statistical tests based on confidence intervals are also used.

Validation Protocol by Type of Method

The information on experimental procedure and acceptance criteria discussed in the previous section is presented in a tabular summary format for each of the types of methods referred to in the guidelines. Please note that the experimental details and typical acceptance criteria are suggestions provided for the purposes of orientation. Therefore, they should be used with caution and may not be appropriate for a particular method. An example of a validation protocol is provided in the case study at the end of this chapter (starting on page 93).

Identification

A generic summary validation protocol is given for identification methods in Table 15.

Table 15 Suggestions for inclusion in a validation protocol for an identification method

Validation characteristic	Experimental details	Typical acceptance criteria
Specificity	Analyse samples which contain the analyte and samples which do not contain the analyte (including samples containing materials which are structurally similar or closely related to the analyte) as per method.	Method must be able to distinguish analyte from materials which are structurally similar or closely related. Positive results obtained for samples containing the analyte and negative results for those not containing the analyte.
	Compare the result obtained by performing the method on the analyte with that obtained for a certified reference standard.	Spectrum/chromatogram/measurement etc. must be the same as that due to a certified reference standard.

Assay

A generic summary validation protocol is given for assay methods in Table 16.

Table 16 Suggestions for inclusion in a validation protocol for an assay method

Validation characteristic	Experimental details	Typical acceptance criteria
Specificity	Analyse samples containing potential interferences, e.g., impurities, excipients. Prepare samples by: Spiking drug substance/drug product with the potentially interfering material. Prepare samples of the potentially interfering materials. Perform stress studies on drug substance/drug product. Chromatographic: Investigate peak purity of analyte peak.	No interference observed for analyte response. Chromatographic: No peaks interfering with analyte peak observed. Minimum resolution between analyte peak and neighbouring peaks = 1.5. No co-elution detected from peak purity investigation.
Linearity	A least 5 concentrations over the range 80-120% of the nominal concentration. An additional solution at 40% is recommended for single point calibration methods. e.g., 40, 80, 90, 100, 110 and 120% nominal	Calibration model valid. Residual plot shows random scatter and no systematic trends. 95% confidence interval of intercept includes zero, or intercept is less than 2% of nominal response.
Range	The range is defined by the results obtained for linearity, accuracy and precision.	Linearity, accuracy and precision demonstrated over the range.
Accuracy	At least 9 determinations over 3 concentration levels e.g., 3 at 80% nominal, 3 at 100%, and 3 at 120%.	Mean recovery within 98-102%. Individual recoveries within 97-103%. Regression analysis of known vs. estimated: Slope within 0.98-1.02 95% confidence interval of slope includes 1
Precision	System precision assessed by 6 replicate measurements/injections	%RSD ≤ 1%
Repeatability	Analysis of 6 preparations from a homogeneous sample at 100% nominal or 3 concentrations/3 levels (use accuracy samples).	%RSD ≤ 2%

Table 16 Suggestions for inclusion in a validation protocol for an assay method

Validation characteristic	Experimental details	Typical acceptance criteria
Intermediate precision	Repeatability samples analysed on different days, instruments, and by different analysts, etc. Use an experimental design.	%RSD ≤ 3%
Reproducibility	Repeatability samples analysed in a different laboratory using design used for intermediate precision.	%RSD ≤ 3%
Robustness	Investigation using an experimental design based on the critical factors identified during method development.	No factor has a significant effect on the assay results or is controlled by an appropriate system suitability test.
Stability	Test solution stability evaluated by comparing stored solutions against results from original analysis. Typical storage conditions include room temperature and refrigerated.	Results for original analysis and reanalysis within 2%.

Drug products may be developed in a series of formulation strengths to provide different dosing options during clinical trials and post approval. The series are usually analysed using the same assay method, the sample preparation is altered to account for the differing amounts of active pharmaceutical ingredient present. During method validation the design of the study needs to ensure that the application of the method to all relevant dosage strengths is covered.

Other types of methods may be validated using a similar approach as for assay methods. Examples are methods for the analysis component of dissolution testing, and methods for testing uniformity of dosage units. Validation of the sample preparation component of dissolution testing involves consideration of the method in question, e.g., the withdrawal and filtration of samples. This topic is discussed by C. C. Chan et al. in 'Analytical Method Validation and Instrument Performance'.[4]

Impurities

A generic summary validation protocol is given for quantitative impurity methods in Table 17.

Table 17 Suggestions for inclusion in a validation protocol for a quantitative impurities method

Validation characteristic	Experimental details	Acceptance criteria
Specificity	Analyse samples containing potential interferences, e.g. impurities, excipients. Prepare samples by: Spiking drug substance/drug product with the potentially interfering material. Prepare samples of the potentially interfering materials. Perform stress studies on drug substance/drug product. Chromatographic: Investigate peak purity of analyte peak.	No interference observed for response due to analyte or impurities of interest. Chromatographic: No peaks interfering with the analyte peak or the peaks due to the impurities of interest observed. Minimum resolution between peaks of interest and neighbouring peaks = 1.5. No co-elution detected from peak purity investigation.
Linearity	A least 5 concentrations over the range QL-120% of the specification limit. e.g., 25, 50, 75, 100 and 120% specification limit of the impurity, the values will depend on the values of the specification limit and the quantitation limit for a particular method.	Calibration model valid. Residual plot shows random scatter and no systematic trends. 95% confidence interval of intercept includes zero, or intercept is less than 2% of nominal response.
Range	The range is defined by the results obtained for linearity, accuracy and precision.	Linearity, accuracy and precision demonstrated over the range.
Accuracy	At least 9 determinations over 3 concentration levels. e.g., 3 at quantitation limit, 3 at 50% of specification limit, and 3 at 120% of specification limit.	Mean recovery within 90-110%. Individual recoveries within 70-130%. Regression analysis of known vs. estimated: Slope within 0.9-1.1 95% confidence interval of slope includes 1
Precision	System precision assessed by 6 replicate measurements/injections.	%RSD ≤ 2%

Table 17 Suggestions for inclusion in a validation protocol for a quantitative impurities method

Validation characteristic	Experimental details	Acceptance criteria
Repeatability	Analysis of 6 preparations from a homogeneous sample at a typical concentration level expected for the impurity or 3 concentrations/3 levels (use accuracy samples).	%RSD ≤ 10%
Intermediate precision	Repeatability samples analysed on different days, instruments, and by different analysts, etc. Use an experimental design.	%RSD ≤ 15%
Reproducibility	Repeatability samples analysed in a different laboratory using design used for intermediate precision.	%RSD ≤ 15%
Detection limit	Prepare 6 samples at or near the expected detection limit (calculate level from required reporting thresholds, i.e., QL/3) by spiking.	~0.01 to 0.02% nominal S/N ~ 3
Quantitation limit	Prepare 6 samples at or near the expected quantitation limit by spiking.	~0.05 to 0.1% nominal S/N ~ 10
Robustness	Investigation using an experimental design based on the critical factors identified during method development.	No factor studied has a significant effect on the assay results, or, factor effects are controlled by an appropriate system suitability test.
Stability	Test solution stability evaluated by comparing stored solutions against results from original analysis. Typical storage conditions include room temperature and refrigerated.	Results for original analysis and reanalysis within 2%.

The validation of quantitative analytical methods for residual solvents follows a similar approach as above where the reporting thresholds for individual solvents are defined in the ICH guideline, 'Impurities: Guideline for Residual solvents'.[5] Limit tests for residual solvents require investigation of specificity and the detection limit only.

Other examples of applications of limit tests are for low-level toxins (highly toxic substances which can be detrimental to the health of the patient at very low concentrations) and polymorphic impurities. A limit test requires less validation than a quantification method and thus is often preferred. More information on the validation of methods for low level toxins is provided by Skett in 'Analysis of Drug Impurities.'[6] Experimental approaches and acceptance criteria similar to that used for impurities may also be applied to methods for water determination.

Bioanalytical methods

A generic summary validation protocol is given for bioanalytical methods in Table 18. This summary is based on the FDA guidelines for bioanalytical method validation.[3]

Table 18 Suggestions for inclusion in a validation protocol for a bioanalytical method

Validation characteristic	Experimental details	Acceptance criteria
Selectivity	Blank samples of the biological matrix from at least 6 sources tested. Samples prepared and tested for all potentially interfering materials, e.g., metabolites.	No interference observed for analyte response. Chromatographic: No peaks interfering with analyte peak observed. Minimum resolution between analyte peak and neighbouring peaks = 1.5. No co-elution detected from peak purity investigation.
Accuracy	Minimum of 5 determinations at 3 concentrations in the expected range.	Mean of each concentration within 15% of the actual value except LLOQ which should not deviate by more than 20%.
Precision	Minimum of five determinations at 3 concentrations in the expected range (use accuracy samples).	%RSD should not exceed 15%, except for LLOQ, where it should not exceed 20%.
Recovery	Compare extracted samples at 3 concentrations (low, medium and high) with unextracted standards that represent 100% recovery.	Recovery need not be 100%, but should be consistent.
Calibration/ Standard Curve	Prepare and analyse 6 to 8 evenly spaced standards covering the required concentration range including the LLOQ, also include a blank sample (matrix only), and a zero sample (matrix and internal standard).	Calibration model valid. Residual plot shows random scatter and no systematic trends. Weighting or complex calibration curve justified.
Lower Limit of Quantification (LLOQ)	At least 5 samples should be used to establish LLOQ.	The response at LLOQ should be at least 5 times the response compared to blank response.

Table 18 Suggestions for inclusion in a validation protocol for a bioanalytical method

Validation characteristic	Experimental details	Acceptance criteria
Stability	Freeze and thaw stability, short-term temperature stability and long-term stability: 3 aliquots of each of high and low concentrations stored and reanalysed alongside control samples. Stock solution stability: Stock solutions stored and reanalysed alongside control samples.	Stored result within 15% of the control. Equivalence shown using statistical test based on confidence interval.

Other methods

For other methods which are not discussed above, the experimental procedure should be based on requirements in the ICH guidelines. The acceptance criteria should be considered on a case by case basis and Table 15 through to Table 18 may provide a starting point for defining suitable values.

Validation of methods in Pharmacopoeias

The tests which are described in individual monographs in the pharmacopoeias have been validated for the substance that the monograph relates to and therefore do not normally require further validation. For methods which are based on separation techniques the system suitability requirements must be fulfilled. However, the FDA regulation 21 CFR 211.194(a)(2) states that 'the suitability of all testing methods used shall be verified under actual conditions of use'.

If tests from pharmacopoeias are applied to non-compendial substances or preparations, then validation is normally required. General tests such as melting point, heavy metals, sulphated ash, pH of solution, etc. may not require further validation but the suitability of the method for the intended use should be demonstrated and may be required in regulatory submissions. A review of the validation of pharmacopoeial methods is given by Miller in 'Method Validation in Pharmaceutical Analysis. A guide to Best Practice'.[7]

Automated methods

The automation of analytical methods is used to increase sample throughput, help ensure data integrity and improve laboratory safety. If a method is considered as being composed of three elements: sample preparation, measurement and data handling then the second and third elements are routinely automated using instrumental analysis. The use of autoinjectors is common in most laboratories and computer controlled systems allow automatic calculation of results. Sample preparation is usually the most labour intensive part of an analysis and automated systems have become popular in laboratories performing pharmaceutical analysis, particularly for quality assurance analysis where the methods are well defined and the sample throughput is high.

The validation of analytical methods which involve automation of measurement requires that the system precision for the measurement and the robustness of the operating parameters is evaluated. For computer controlled data handling the software need to be fully validated before it can be used in an analysis. For automated sample preparation there are a number of possible impacts on the approach to method validation and these depend largely on the nature of the method. If the method was originally developed as a manual method, then the validation will consist of demonstrating equivalency between the automated and manual methods. If the automated method was developed without a manual method, then the validation will follow the ICH guidelines. However, the design of the validation will need to consider the unique attributes related to the way in which the sample handling system. Also, the equipment qualification, similar to an analytical instrument, should be performed prior to performing validation studies. Useful guidance regarding development and validation of automated methods is provided by Incledon and Ham in 'Analytical Method Validation and Instrument Performance Qualification'.[8]

Validation by Phase of Development

An analytical method which is used in pharmaceutical analysis should always be validated to ensure that the results generated are trustworthy. These results may be used to make critical decisions during the drug development process relating to issues such as safety of the drug, the synthetic route and the manufacturing process. However, performing formal validation studies as per ICH guidelines requires considerable resources. In the early stages of the development of a drug the investment required for formal validation may not be desirable due to a number of reasons including:

- The development and optimisation of the analytical methods is ongoing.
- The development of the synthetic route for the drug substance is ongoing.
- The development of the formulation of the drug product is ongoing.
- The drug may not progress into later stages of development.

The ICH guidelines apply to 'validation of the analytical procedures included as part of registration applications submitted within the EC, Japan and USA'[1], and thus do not formally apply to the early phases of drug development. The guidelines from the FDA regarding INDs for phase 2 and phase 3 studies[9] require 'appropriate' validation data for methods which are not from a pharmacopoeia or official reference standard.

The consequence of this is that it is common for pharmaceutical companies to use a phased approach to analytical method validation studies. Validation performed for early phase drugs tends to be less extensive than that performed for late stage drugs. However, the objective of validation still applies, i.e., to demonstrate that a method is suitable for its intended purpose. The following list provides suggestions for validation of early phase drugs:

- A formal validation protocol is not yet mandatory. Internal guidelines, or a standard operating procedure (SOP), may be used to summarise the general validation requirements. This may be referenced rather than producing time consuming documentation.
- The extent of testing and the number of replications may be reduced.
- Wider acceptance criteria may be adequate in early phases of development.
- Specificity and evaluation of the quantitation limit are the primary characteristics to ensure that assay and impurity methods meet their intended purposes of potency and safety.
- A second method to evaluate accuracy is unlikely to exist in the early phase of development, therefore accuracy may be inferred from the results of precision, linearity and specificity.
- For precision testing synthetic mixtures of drug substance and placebo may be used, rather than authentic samples, thus enabling the combination of accuracy, precision and potentially linearity at the same time.
- Formal intermediate precision experiments are not yet needed but if different laboratories need to operate the method, then the handover will require suitable validation.
- The evaluation of the detection limit for impurity methods may be delayed.
- Formal robustness testing is not yet required. Robustness studies associated with method development are likely to be ongoing at this stage.
- The validation report may be presented in a simplified tabular format, together with the conclusions. This type of summary report fulfils the expectations of the regulatory authorities, e.g., phase 2 and phase 3 INDs

Further information regarding validation of analytical methods by phase of development is available from Bloch in "Method Validation in Pharmaceutical Analysis. A guide to Best Practice'[10] and also from Boudreau et al. in a paper developed from a PhRMA 2003 workshop.[11]

Validation Software

There are a number of different types of computer software available to assist in the process of performing analytical method validation. These range from statistics packages[12,13] which may be used to perform the necessary statistical calculations associated with method validation such as linear regression, to experimental design software[14,15] used for robustness and intermediate precision studies, to software specifically designed to enable validation of analytical methods.[16-23] There are different configurations for the latter type of validation software, most commonly they are applied to chromatographic methods and are incorporated into chromatography data systems. A consideration common to all method validation software, whether it is a sophisticated method validation software package, or a purpose built spreadsheet, is that the software itself must be suitably validated. Useful guidance on the validation of spreadsheets is provided by Phan.[24]

Planning and Execution of the Validation Study

The protocol defines the experiments which are required for the validation study. Careful planning and execution must be applied to perform these experiments in the most efficient way. Consideration needs to be given to the timing and resources needed for each of the required experiments and also to the most appropriate order in which they should be performed. Experiments should be planned in such a way as to make the best use of available time and resources, but a realistic and pragmatic approach is essential to ensure that too much is not attempted in one experiment. An experiment which aims to achieve multiple outcomes can lead to confusing datasets which are difficult to interpret and can ultimately result in reanalysis.

The process can be split into three steps, these are:

1. Review the timing and resources required for each validation characteristic.

2. Determine the most efficient order of experiments and combine experiments where appropriate.

3. Execute the study by securing all necessary resources and commencing the experiments.

Planning and execution is considered in the case study at the end of this chapter (see page 99).

Review the timing and resources required

The nature of both the type of method and the type of analytical technique will influence the timing and resources required for the experiments associated with each validation characteristic. The review of the validation protocol to evaluate the timing and the resources which are required should include consideration of the following:

- The time taken to perform the analytical method should be evaluated. For example, in the case of a chromatographic analysis the run time for each injection will impact the number of samples which may be analysed per day.

- Ensure that all equipment and reagents which are required to perform the analytical method are available in sufficient quantity to perform the validation study, e.g., order chromatographic columns, reference standards, etc. Similarly, if a supporting analytical method is being used during the validation study, then the equipment and reagents for this method need to be available.

- If specificity testing involves a forced degradation study, then this will take time both to set up and to run. The time required for sufficient degradation may be difficult to predict. An efficient approach may be to set up the study prior to other validation testing and arrange the analysis of the stressed samples to correspond to the specified degradation times. Extra time should be allowed for the event of insufficient degradation at the specified times.

- Samples for specificity testing relating to chromatographic methods may be tested using diode array detection or mass spectrometry. The instrumentation required for this analysis may need to be secured.

- The total number of analytical samples that require testing for each validation characteristic should be determined so that the time which is required to perform the study can be estimated.

Determine the most efficient order of experiments

The order in which to perform validation experiments may be defined by the test solutions which are required for the testing. Where possible experiments are combined, and the preparation of test solutions is minimised. The following should be considered during the planning of experiments:

- Test solutions which are prepared for testing one validation characteristic may be used for another. Examples are solution stability and robustness studies, where the samples prepared for precision testing may be used. Also test solutions prepared for linearity and accuracy may overlap. In situations where authentic samples are not possible for precision testing, accuracy and precision can be determined using the same test solutions.

- Combining experiments so that they are analysed together makes good use of available time and resource. The effectiveness of this approach will depend on the times involved in the analysis for both sample preparation and measurement of the test solution.

- For linearity studies where two ranges are being evaluated and then later compared, e.g., combined assay and impurity methods by HPLC, it is advisable to perform the linearity experiments at the same time and on the same equipment.

- Formal validation studies performed for robustness testing which involve design of experiments require a number of experiments which ideally are performed in a random order. Since these studies involve changing the method parameters they cannot be used to produce data for other validation characteristics.

- Intermediate precision studies should be performed using two operators and each time the method is applied by each operator all test solutions and reagents such as diluent, mobile phase etc. should be prepared from scratch.

Execute the study

When the validation protocol is complete and the required experiments have been planned, then the study can commence. Prior to performing any laboratory work the following should be considered:

- The raw data for the validation experiments should be recorded using a laboratory notebook or equivalent in GMP/GLP (as appropriate) compliant manner.

- A well characterised certified reference standard should be used for the study.

- The equipment used during the study should be appropriately qualified.

Documentation

The format and the approval of the documentation for the validation protocol will depend on the requirements of the in-house validation policy for a particular laboratory. This in turn may depend on the stage of development for the drug being studied. Validation studies performed for drugs in the early stages of development may not need as much documentation as those in late stages when a formal validation protocol is required by GMP guidelines. A tabular summary of the validation protocol may be sufficient for early phase studies. A formal validation protocol for late stage development will include a detailed account of the validation characteristics to be studied, together with the acceptance criteria and the experimental procedure to be applied.

Case Study

An example of a HPLC analytical method which is used for the analysis of 'MiracleCure' 25mg tablets is presented in Figure 12. This method is used to assay the drug product and also for the determination of degradation products. A suggested validation protocol for this method is presented in Table 19.

Figure 12 An example of a HPLC analytical method

Analytical Method for Determination of the Assay and Degradation Products of 'MiracleCure' 25 mg Tablets by HPLC

Chromatographic parameters

Instruments	HPLC system with a UV detector			*Flow rate*	1 mL/min
Column	Symmetry C18, 10cm, 4.6mm i.d., 3.5µ			*Injection volume*	5µL
Mobile Phase A	25mM Potassium phosphate, pH 3			*Temperature*	30°C
Mobile Phase B	Acetonitrile			*Wavelength*	240nm
Gradient	Time	%A	%B		
	0	80	20		
	40	30	70		
	40.1	80	20		
	50	80	20		

Preparation of test solutions

Diluent 50/50 acetonitrile/water

SST solution Weigh approximately 2.5mg of Impurity X analytical reference standard into a 50mL volumetric flask and make to volume with diluent. Call this solution X. Weigh accurately approximately 25mg of 'MiracleCure' analytical reference standard into a 100mL volumetric flask, then pipette 1mL of solution X into the same flask. Make to volume with diluent.

Calibration standard Weigh accurately approximately 25mg of 'MiracleCure' analytical reference standard into a 100mL volumetric flask and make to volume with diluent.

Standard check Weigh accurately approximately 25mg of 'MiracleCure' analytical reference standard into a 100mL volumetric flask and make to volume with diluent.

LOQ solution Pipette 1mL of the calibration standard in to a 100mL volumetric flask and make to volume with diluent to obtain a 1% solution. Pipette 1mL of the 1% solution into a 20mL volumetric flask and make to volume with diluent to obtain the LOQ solution.

Sample Weigh accurately 20 'MiracleCure' 25 mg tablets. Mill tablets for 5 minutes to obtain a homogeneous powder. Weigh accurately approximately 250mg of the powder into a 100mL volumetric flask and add approximately 70 mL of diluent. Sonicate for 10 minutes and allow to equilibrate to room temperature. Make to volume with diluent. Prepare in duplicate.

Procedure

Inject the SST solution, the diluent, The LOQ solution, the standard check, the calibration standard and the samples. 6 Injections of the calibration standard are required for repeatability, thereafter the calibration standard should be injected every four hours between samples and the last injection of the analysis should be of calibration standard.

Figure 12 continued

The peak due to 'MiracleCure' in the SST solution should elute at a retention time in the range of 9 to 12 minutes. The peak due to Impurity X in the SST solution should elute at a relative retention time of 1.21. The resolution between the peak due to 'MiracleCure' and the peak due to Impurity X in the SST solution should be > 2.0. The tailing factor for the peak due to 'MiracleCure' in the SST solution should be <2. The %RSD for 6 injections of standard should be not more than 1.0%. There should be no interfering peaks in the diluent injection.

Use all the injections of calibration standard to construct a calibration curve for the peak due to 'MiracleCure'.
Calculate the recovery of the standard check solution, the recovery of the peak due to 'MiracleCure' in the standard check should be within the range 99.0 to 101.0%.
Calculate the recovery of the LOQ solution, the recovery of the peak due to 'MiracleCure' in the LOQ should be within the range 0.04 to 0.06%.
Calculate the assay for 'MiracleCure' in the sample.
Calculate each of the degradation products in the sample disregarding peaks below 0.05% w/w.

Table of peaks with expected retention times and relative retention times:

Name	Retention time (minutes)	Relative Retention Time (RRT)
MiracleCure	10.5	1.00
Impurity X	12.7	1.21
Degradation Product 1	15.4	1.4
Degradation Product 2	37.2	3.54

Calculations
The calculation for %RSD, resolution and tailing factor are as detailed in USP 30 <621>.

Recovery of standard check solution =

$$\frac{A_{std\ check}}{A_{cal\ std}} \times \frac{W_{cal\ std}}{W_{std\ check}} \times 100$$

Where:
$A_{std\ check}$ = Area of peak due to 'MiracleCure' in standard check solution injection
$A_{cal\ std}$ = Average area of peak due to 'MiracleCure' in calibration standard solution injections
$W_{std\ check}$ = Amount of 'MiracleCure' in standard check solution (mg)
$W_{cal\ std}$ = Amount of 'MiracleCure' in calibration standard solution (mg)

Recovery of LOQ solution =

$$\frac{A_{LOQ}}{A_{cal\ std}} \times 100$$

Where:
A_{LOQ} = Area of peak due to 'MiracleCure' in LOQ solution injection

Assay (mg/tablet) =

$$\frac{A_{sam}}{A_{cal\ std}} \times \frac{W_{cal\ std}}{W_{sam}} \times \frac{W_{20}}{20}$$

Where:
A_{sam} = Area of peak due to 'MiracleCure' in the sample solution injection
W_{sam} = Amount of 'MiracleCure' milled 25 mg tablet powder in sample solution (mg)
W_{20} = Mass of 20 'MiracleCure' 25mg tablets (mg)

Figure 12 continued

Assay (% label claim) =

$$\frac{\text{Assay (mg/tablet)}}{25} \times 100$$

Assay should be determined for each duplicate and the mean of the results reported.

Degradation Products (% with respect to 'MiracleCure') =

$$\frac{A_{deg}}{A_{sam}} \times 100$$

Where:

A_{deg} = Area of peak due to the degradation product in the sample solution injection

Each degradation product above the disregard limit should be quantified for each duplicate sample preparation and the mean of the results reported.

Further information regarding the method

MiracleCure 25mg tablets are about to be used in the first phase 3 clinical trials. Degradation Product 1 (DP 1) and Degradation Product 2 (DP 2) are the major degradation products observed for the formulation in stability trials and stress studies. DP 1 has been isolated from a forced degradation sample and thus a sample is available for the validation study. No isolated sample is available for DP 2. The probable specification limits (SL) are 0.5%w/w for DP 1 and 0.2%w/w for DP 2, both expressed with respect to the active pharmaceutical ingredient. Impurity X is a process impurity and is controlled in the drug substance.

Table 19 Suggested validation protocol for the analytical method in Figure 12

Validation characteristic	Experimental details	Acceptance criteria
Specificity	Prepare and analyse samples of the following: 1. MiracleCure, at the nominal concentration (0.25 mg/mL). 2. Impurity X, spiked into a solution of MiracleCure at a level of ~0.2% w/w. 3. DP 1, spiked into a solution of MiracleCure at a level of ~0.2% w/w. 4. Samples from stress studies containing DP 2 present at a level of ~0.1% w/w. 5. Excipients used in MiracleCure 25mg tablets at the concentration typical of the method.	No peaks interfering with the peaks due to MiracleCure, Impurity X, DP 1 or DP 2. Minimum resolution between peaks of interest and neighbouring peaks = 2.

Table 19 Suggested validation protocol for the analytical method in Figure 12

Validation characteristic	Experimental details	Acceptance criteria
Linearity – Assay	Prepare and analyse (single injections) linearity standards of MiracleCure prepared at the following concentrations in the presence of excipients at a typical sample level: 40% nominal, 0.1 mg/mL 80% nominal, 0.2 mg/mL 90% nominal, 0.225 mg/mL 100% nominal, 0.25 mg/mL 110% nominal, 0.275 mg/mL 120% nominal, 0.3 mg/mL Should be performed at the same time as the linearity study for DP 2 and unknown degradation products.	Calibration model valid. r > 0.999 Residual plot shows random scatter and no systematic trends. 95% confidence interval of intercept includes zero, or intercept is less than 2% of nominal response.
Linearity – DP 1	Prepare and analyse (triplicate injections) linearity standards of DP 1 prepared at the following concentrations in the presence of MiracleCure and excipients at a typical sample level: 0.05% nominal, 0.125 µg/mL = QL 0.1% nominal, 0.25 µg/mL 0.25% nominal, 0.625 µg/mL = 50% SL 0.4% nominal, 1.0 µg/mL 0.5% nominal, 1.25 µg/mL = SL 0.6% nominal, 1.5 µg/mL	Calibration model valid. r > 0.99 Residual plot shows random scatter and no systematic trends. 95% confidence interval of intercept includes zero, or intercept is less than 2% of nominal response.
Linearity – DP 2 & unknown degradation products	Prepare and analyse (triplicate injections) linearity standards of MiracleCure prepared at the following concentrations in the presence of excipients at a typical sample level: 0.05% nominal, 0.125 µg/mL = QL 0.075% nominal, 0.1875 µg/mL 0.1% nominal, 0.25 µg/mL = 50% SL 0.15% nominal, 0.375 µg/mL 0.2% nominal, 0.5 µg/mL = SL 0.24% nominal, 0.6 µg/mL	Calibration model valid. r > 0.99 Residual plot shows random scatter and no systematic trends. 95% confidence interval of intercept includes zero, or intercept is less than 2% of nominal response. Slope is within 5% of slope obtained for assay linearity.

Table 19 Suggested validation protocol for the analytical method in Figure 12

Validation characteristic	Experimental details	Acceptance criteria
Range	The range is defined by the results obtained for linearity, accuracy and precision.	Linearity, accuracy and precision demonstrated over the range of QL to 120% nominal concentration, or 0.000125 to 0.3 mg/mL.
Accuracy – Assay	Prepare and analyse samples of MiracleCure at the following concentrations in the presence of excipients at a typical sample level: 3 replicates at 80% nominal, 0.2 mg/mL 3 replicates at 100% nominal, 0.25 mg/mL 3 replicates at 120% nominal, 0.3 mg/mL	Mean recovery within 98-102%. Individual recoveries within 97-103%. Regression analysis of known vs. estimated: Slope within 0.98-1.02 95% confidence interval of slope includes 1
Accuracy – DP 1	Prepare and analyse samples of DP 1 at the following concentrations in the presence of MiracleCure and excipients at a typical sample level: 3 replicates at QL, 0.125 μg/mL 3 replicates at 50% SL, 0.625 μg/mL 3 replicates at 120% SL, 1.5 μg/mL	Mean recovery within 90-110%. Individual recoveries within 70-130%. Regression analysis of known vs. estimated: Slope within 0.9-1.1 95% confidence interval of slope includes 1
Precision	System precision for assay assessed by 6 replicate injections of calibration standard during precision analysis.	%RSD for 6 injections ≤ 1%
	System precision for degradation products assessed by 6 replicate injections of the LOQ solution during precision analysis.	%RSD for 6 injections ≤ 10%
Repeatability – Assay	Prepare and analyse 6 preparations from a representative batch of milled MiracleCure 25mg tablets.	%RSD for 6 assay results ≤ 2%
Repeatability – DP 1	Prepare and analyse 6 preparations from a stressed batch of milled MiracleCure 25mg tablets which contains DP 1 at a level of approximately 0.2%w/w w.r.t. MiracleCure.	%RSD for 6 DP 1 results ≤ 10%
Repeatability – DP 2	Prepare and analyse 6 preparations from a stressed batch of milled MiracleCure 25mg tablets which contains DP 2 at a level of approximately 0.1%w/w w.r.t. MiracleCure.	%RSD for 6 DP 2 results ≤ 10%

Table 19 Suggested validation protocol for the analytical method in Figure 12

Validation characteristic	Experimental details	Acceptance criteria
Intermediate precision	Repeatability samples for Assay, DP 1 and DP 2 analysed using the following experimental design: Run 1 / Run 1 / Run 3 / Run 4 Day 1 / Day 1 / Day 2 / Day 2 Anal 1 / Anal 2 / Anal 1 / Anal 2 Inst 2 / Inst 1 / Inst 1 / Inst 2 Where: Anal = Analyst, Inst = Instrument	Assay: %RSD ≤ 3% Degradation products: %RSD ≤ 15%
Detection limit – DP 1	Prepare and analyse 6 samples of DP 1 at a concentration of 0.02% nominal (0.05 µg/mL) in the presence of MiracleCure and excipients at a typical sample level.	DL is 0.02% nominal S/N ~ 3
Detection limit – DP 2 and unknown degradation products	Prepare and analyse 6 samples of MiracleCure at a concentration of 0.02% nominal (0.05 µg/mL) in the presence of excipients at a typical sample level.	DL is 0.02% nominal S/N ~ 3
Quantitation limit – DP 1	Prepare and analyse 6 samples of DP 1 at a concentration of 0.05% nominal (0.125 µg/mL) in the presence of MiracleCure and excipients at a typical sample level.	QL is 0.05% nominal S/N ~ 10
Quantitation limit – DP 2 and unknown degradation products	Prepare and analyse 6 samples of MiracleCure at a concentration of 0.05% nominal (0.125 µg/mL) in the presence of excipients at a typical sample level.	QL is 0.05% nominal S/N ~ 10

Table 19 Suggested validation protocol for the analytical method in Figure 12

<table>
<tr><th>Validation characteristic</th><th>Experimental details</th><th>Acceptance criteria</th></tr>
<tr><td>Robustness</td><td>Perform a robustness study using Plackett-Burman experimental design on the following factors and levels using a sample containing MiracleCure and associated impurities:
<table>
<tr><th>Factor</th><th>-1</th><th>Nom</th><th>+1</th></tr>
<tr><td>%B in MP (start)</td><td>19</td><td>20</td><td>21</td></tr>
<tr><td>%B in MP (end)</td><td>69</td><td>70</td><td>71</td></tr>
<tr><td>Buffer conc</td><td>24</td><td>25</td><td>26</td></tr>
<tr><td>Buffer pH</td><td>2.7</td><td>3.0</td><td>3.3</td></tr>
<tr><td>Flow Rate</td><td>0.9</td><td>1.0</td><td>1.1</td></tr>
<tr><td>Temperature</td><td>27</td><td>30</td><td>33</td></tr>
<tr><td>Wavelength</td><td>240</td><td>245</td><td>245</td></tr>
</table>
Nom = nominal level for the factor
Responses to be studied:
Results for MiracleCure, DP 1 and DP 2. System suitability results from the criteria in the method.</td><td>No factor studied has a significant effect on the assay results, or, factor effects are controlled by an appropriate system suitability test.</td></tr>
<tr><td>Stability</td><td>Store 3 of the preparations from repeatability study at room temperature protected from light, and 3 refrigerated. Reanalyse after 48 hours and again after 7 days. Compare with results of original analysis.</td><td>Results for MiracleCure in the original analysis and reanalysis within 2%.
Results for DP 1 and DP 2 in the original analysis and reanalysis within 10%.</td></tr>
</table>

Planning and execution of the study

To plan the experiments associated with this validation study the protocol was reviewed and the timing and resources required for each validation characteristic assessed. The long analysis times for the experiments for each validation characteristic are due to the long run time of the HPLC method (50 minutes). This limits any benefits which may be derived from the combination of experiments. It is estimated that the validation study will require a period of approximately 6 weeks for completion.

Specificity

A forced degradation study has been performed previously for this analytical method and thus is not required for this validation. Five samples have been identified which will be used to demonstrate specificity. With the exception of the test solution containing excipients, the test solutions to be tested will be prepared for other validation characteristics. Test solution 1 is the calibration standard and 2 is the SST solution, both prepared during precision testing. Test solution 3 is prepared during accuracy testing and 4 during precision testing. A HPLC system which has diode array detection and mass spectrometry capabilities will be used for the analysis to enable peak purity investigations.

Linearity

The linearity studies for the assay and impurities involve the preparation of 18 linearity test solutions: 6 for the assay method, 6 for impurity DP1, and 6 for impurity DP2 and unknown degradation products. The test solutions will be prepared by dilution from a stock solution for each linearity test using a diluent containing the excipients at the appropriate concentration. The assay test solutions are injected singly but the impurity solutions are injected in triplicate resulting in a total of 42 injections. The runtime for a single injection on the HPLC instrumentation is 50 minutes and therefore the measurement of the test solutions will require approximately 36 hours.

Accuracy

The accuracy studies for the assay and impurities involve the preparation of 18 accuracy test solutions: 9 for the assay method, and 9 for impurity DP1. These will be prepared using individual weighings to which an appropriate amount of excipient blend will be added before dissolving in diluent. The accuracy test solutions will be analysed as per method, i.e., 4 test solutions injected between injections of calibration standard. The total number of injections is 32 and the total runtime will be approximately 27 hours.

Precision

The repeatability studies for the assay and impurities involve the preparation of 18 precision test solutions: 6 for the assay method, 6 for impurity DP1, and 6 for impurity DP2. The test solutions are prepared and analysed as per method. The total number of injections is 37 (including 6 injections of the LOQ solution for system precision) and the total runtime will be approximately 31 hours.

The intermediate precision studies require repeating the above testing another three times. This is performed once by the same operator and twice by another operator.

Detection limit

The detection limit study involves the preparation of 12 test solutions: 6 for impurity DP1, and 6 for impurity DP2 and unknown degradation products. The test solutions are prepared in the same way as the accuracy test solutions and analysed as per method. The total number of injections is 25 and the total runtime will be approximately 21 hours.

Quantitation limit

The detection limit study involves the preparation of 12 test solutions: 6 for impurity DP1, and 6 for impurity DP2 and unknown degradation products. The test solutions are prepared in the same way as the accuracy test solutions and analysed as per method. The total number of injections is 25 and the total runtime will be approximately 21 hours.

Robustness

The robustness study involves 8 separate experiments in which the method parameters are altered differently for each one. A test solution containing the active pharmaceutical ingredient, MiracleCure, and the impurities of interest, Impurity X, DP1 and DP2, is prepared by spiking and is analysed as per method (with the exception of the alteration defined by the study) for each experiment. The total number of injections required is 13, which will have a total runtime of approximately 11 hours.

Stability

The study to investigate the stability of test solutions will be performed using the test solutions prepared for the repeatability study. The repeatability results will be used as the original analysis and the reanalysis will be performed at 48 hours and at 7 days using freshly prepared calibration solutions each time. The total number of injections for the reanalysis is 32 and the total runtime will be approximately 27 hours.

Summary

1. The validation protocol details the design of the validation study for an analytical method.
2. The validation characteristics which should be studied for a particular type of method are defined in the ICH guidelines and supplemented by FDA guidelines.
3. The validation protocol contains the requirements for the experimental procedure to be followed for the validation study and the acceptance criteria which should be applied to the results.
4. The phase of development of the drug under investigation may influence the extent of the validation performed, less extensive validation is usually performed for early phase projects.
5. Software which may be used during analytical method validation includes: statistics packages, experimental design packages, and analytical method validation packages. All software should be validated.
6. The experimental work associated with analytical method validation should be performed in a GMP/GLP compliant manner, using certified reference standards and qualified equipment.

References

1. International Conference on Harmonisation (ICH) of Technical Requirements for Registration of Pharmaceuticals for Human Use, Topic Q2 (R1): **Validation of Analytical Procedures: Text and Methodology**, 2005, www.ich.org.

2. Draft guidance for industry: **Analytical Procedures and Methods Validation**, US Food and Drug Administration, Center for Drugs and Biologics, Department of Health and Human Services, 2000.

3. Guidance for Industry: **Bioanalytical Method Validation**, US Food and Drug Administration, Center for Drugs and Biologics, Department of Health and Human Services, 2001.

4. C. C. Chan, N. Pearson, A. Rebelo-Cameirao '**Dissolution Method Validation**', in 'Analytical Method Validation and Instrument Performance Verification', Eds. C. C. Chan, H. Lam, Y. C. Lee, X. Zhang, Wiley, 2004, p51-66.

5. International Conference on Harmonisation (ICH) of Technical Requirements for Registration of Pharmaceuticals for Human Use, Topic Q3C(R3): **Impurities: Guidelines for Residual Solvents**, 2002, www.ich.org.

6. P. Skett, '**Low-level measurement of potent toxins**', in 'Analysis of Drug Impurities', Eds. R. J. Smith, M. L. Webb, Blackwell Publishing, 2007, p82-123.

7. J. H. Miller, '**Validation of Pharmacopoeial Methods**', in 'Method Validation in Pharmaceutical Analysis, a Guide to Best Practice', Eds. J. Ermer, J. H. Miller, Wiley, 2005, 301-336.

8. C. Incledon, H. Lam, '**Development and Validation of Automated Methods**', in 'Analytical Method Validation and Instrument Performance Verification', Eds. C. C. Chan, H. Lam, Y. C. Lee, X. Zhang, Wiley, 2004, p67-83.

9. Guidance for Industry: **INDs for Phase 2 and Phase 3 Studies**, Chemistry, Manufacturing, and Controls information, US Department of Health and Human Services, Food and Drug Administration, Center for Drug Evaluation and Research (CDER), 2003.

10. M. Bloch, '**Validation During Drug Product Development – Considerations as a Function of the Stage of Drug Development**', in 'Method Validation in Pharmaceutical Analysis, a Guide to Best Practice', Eds. J. Ermer, J. H. Miller, Wiley, 2005, p243-264.

11. S. P. Boudreau, J. S. McElvain, L. D. Martin, T. Dowling, S. M. Fields, *Pharm. Technol.*, **28** (11), 54-66, 2004, '**Method Validation by Phase of Development – An Acceptable Analytical Practice**'.

12. Software 'Minitab', Minitab Inc., www.minitab.com.

13. Software 'Microsoft Excel', Microsoft Corporation, www.office.microsoft.com.

14. Software 'MODDE', Umetrics, www.umetrics.com.

15. Software 'The Unscrambler', CAMO Software, www.camo.com.

16. Software 'MVA 2.0 – Method Validation in Analytics', www.novia.de.

17. Software 'Validation Manager', VWR International, www.vwr.com.

18. Software 'Fusion AE', S-Matrix Corporation, www.smatrix.com.

19. Software 'Chemstation Plus Method Validation Pack', Agilent Technologies, www.agilent.com.

20. Software 'Empower 2 Method Validation Manager', Waters, www.waters.com.

21. J. Flarakos, *LC•GC*, **19** (3), 304-310, 2001, '**Analytical Validation Software**'.

22. P. Lukulay, R. Versepur, *Pharm. Technol.*, **30** (5), 2005, '**Automating HPLC and GC Analytical Method Validation**'.

23. P. Lukulay, J. Morgado, *LC•GC*, **24** (2), 2006, '**Automation of the Chromatographic Analytical Method Validation Process**'.

24. T. T. Phan, *Pharm. Technol.*, 50-62, Jan 2003, '**Technical Considerations for the validation of electronic spreadsheets for complying with 21 CFR Part 11**'.

Further reading

J. Ermer, J. H. Miller, '**Method Validation in Pharmaceutical Analysis, a Guide to Best Practice**', Wiley, 2005.

D. M. Bliesner, '**Validating Chromatographic Methods: A practical guide**', Wiley, 2006.

C. C. Chan, Y. C. Lee, H. Lam, X. Zhang, '**Analytical Method Validation and Instrument Performance Verification**', Wiley, 2004.

Notes

The Validation Report

When the experiments detailed in the validation protocol have been performed the next stage of the validation study is the interpretation of the results. The calculations and statistics associated with each validation characteristic are carried out and the results obtained are assessed against the acceptance criteria to decide if the method meets the validation requirements.

Contents of the Validation Report

The validation report details the results of the validation study. Its purpose is to provide information on which characteristics were tested during the study, the results obtained, and the interpretation of those results. Typical information in a validation report may include:

- Details of the validation protocol.
- Details regarding the analytical method (or methods) validated.
- The validation characteristics which were investigated for the method(s).
- The results which were calculated for each validation characteristic.
- A discussion of the interpretation of the results and how they compare to the acceptance criteria.
- Any relevant validation information which was obtained during method development, e.g., solution stability, robustness, stress studies, etc.
- Details of the reference materials used in the validation study.
- Details relating to batch numbers, etc. of the materials used in the validation study.
- Details of the equipment used in the study, e.g., identifiers and qualification details.
- References to the laboratory notebooks, or equivalent used to record the raw data obtained during the validation study.

- Details of responsibilities and required signatures.

Statistics in Analytical Method Validation

The statistics required for the interpretation of validation results include: the calculation of the mean, standard deviation, confidence intervals and relative standard deviation for data sets obtained; regression analysis for evaluation of linearity and accuracy (these were discussed in Chapter 2); comparative studies; and assessment of the significance of outliers. Validated statistical software packages[2,3] are normally used for the calculation of all statistics associated with analytical method validation.

Statistical significance

A statistically significant difference means that there is statistical evidence that a difference exists which is unlikely to have occurred by chance. It does not mean that the difference is necessarily large or important and thus may not be of practical relevance. Statistics are very useful to support analytical method validation, but the results should be used with caution. Sound scientific judgement is required for interpretation of statistical results.

Comparative studies

Examples of comparative studies which could be performed during method validation are:

- Specificity and accuracy, to compare the results obtained using the method being validated against a second well characterised method.
- Precision, to compare the results obtained during intermediate precision and reproducibility studies.
- Robustness, to compare the effects due to the factors under evaluation.
- Stability of solutions, to compare the results obtained from solution which have been stored against the original results or results obtained for a freshly prepared sample.

Student's t-test

Student's t-tests are one of the most commonly used statistical significance tests applied to small data sets. A t-test may be used to compare two sets of data which are each characterised by their mean, standard deviation and number of data points, provided that the distribution of the data can be assumed to be normal. The outcome of a t-test is that the null hypothesis is accepted or rejected. The null hypothesis is that any differences between the data sets are due to random and not systematic errors, i.e., both methods give the same results, or both samples contain the same amount of analyte.

There are different types of t-tests which can be applied, the choice depends on the nature of the data. This can be:

1. Dependent, or paired: where data sets typically consist of a matched data set or one group that has been tested twice. An example is the stability of test solutions where the same solution is analysed at different times and the results for the original analysis and those for the reanalysis are compared.

2. Independent: where data sets typically consist of two groups with no relationship. This may be split further into:

   i. Equal variance: Data sets which have similar standard deviations. Examples are intermediate precision and reproducibility where the data is generated using the same samples and method but under different conditions, e.g. analysts, instruments, laboratories, etc.

   ii. Unequal variance: Data sets which have dissimilar standard deviations. Examples are specificity and accuracy investigations where two methods are used to analyse the data and thus the variability may be different for each method.

The test is performed by the calculation of a t value which is compared to the Student's t-distribution table of values (available in statistic textbooks and introduced previously in Chapter 2, see Table 4). The t value which is associated with the degrees of freedom of the combined data sets and a particular confidence level (usually 95%) can be obtained from the table. If the calculated t value is above the value from the table, then the null hypothesis (no difference) is rejected and it is concluded that there is a statistical difference.

The equation which is used for the calculation of the t-value (and also the degrees of freedom for the data sets) is determined by the type of t-test being applied, i.e., paired, equal variance or unequal variance. The calculations are based on the mean, the standard deviation and the number of data points for each data. These equations are available in textbooks on statistics but in practice the test is usually performed using a software package for statistics such as Minitab[2] or Excel.[3]

For most purposes a two tailed t-test is applied, this means that the null hypothesis is rejected if either an unusually large or an unusually small value of the test statistic is obtained, the rejection region is divided between the two tails. A one tailed t-test is used to show that a particular mean is larger or smaller than the other, in this case the rejection region is entirely in one tail.

More information on statistical tests for comparative studies may be found in statistical textbooks (see further reading at the end of this chapter), and also in the chapter on simple comparative studies in the book, 'Statistics for the quality control laboratory'[4] by Mullins.

Outliers

Occasionally in validation studies, as in any analysis, outliers may be observed in data sets of results. This is where a value is present in the data set which differs considerably from the majority of the other results. For example, the following data set was obtained for a precision study:

25.4, 25.3, **27.5**, 24.5, 24.7, 25.6

The value 27.5 is much higher than the one nearest to it at 25.6 and it is suspected that it may be an outlier. The mean and standard deviation calculated for the data set including and excluding the suspected outlier are presented in Table 20. The mean does not appear to be affected substantially by the suspected outlier but if the suspect value is included the %RSD does not comply with the acceptance criterion of %RSD $\leq$ 2% for precision. If it is excluded the precision complies with the acceptance criterion.

Table 20 Mean and %RSD for the data set containing a suspected outlier

Sample	Results (mg/tablet)
1	25.4
2	25.3
3	27.5
4	24.5
5	24.7
6	25.6
Mean (inc. suspected outlier)	25.5
%RSD (inc. suspected outlier)	3.82
Mean (ex. suspected outlier)	25.1
%RSD (ex. suspected outlier)	1.89

Statistical tests can be performed which will provide confidence in the characterisation of a data point as an outlier. However, a decision still has to be made about whether to exclude the data point from the results. Some statisticians object to the rejection of any data from small size data sample, unless it is known that something went wrong during the measurement of that data. Rejection of data during validation studies must be very carefully considered, statistical evidence alone may not be enough to justify the rejection of data.

The most popular statistical test applied to detect suspect outliers in the results from chemical analysis is Dixon's Q-test. One (and only one) observation from a small set of replicate observations (typically 3 to 10) can be examined. The test assumes a normal distribution of the data. The null hypothesis for this test is that there is no significant difference between the suspect value and the rest of the values, any differences must be attributed to random errors.

The test is applied as follows:

1. The values comprising the data set are arranged in ascending order:

   $X_1 < X_2 < ...... < X_n$

   e.g., 24.5, 24.7, 25.3, 25.4, 25.6, 27.5

2. The experimental Q-value is calculated, defined as the ratio of the difference between the suspect value and the nearest value to it, to the range of the data.

   If the suspected outlier is a low value:

   $$Q_{exp} = \frac{X_2 - X_1}{X_n - X_1}$$

   If the suspected outlier is a high value:

   $$Q_{exp} = \frac{X_n - X_{n-1}}{X_n - X_1}$$

   e.g., $Q_{exp} = (27.5 - 25.6)/(27.5 - 24.5) = 0.633$

3. The value of Q_{exp} is compared to a critical Q-value (Q_{crit}). Refer to Table 21 for critical values of Q for Dixon's test, from work by Rorabacher.[5] The value of Q_{crit} corresponding to the confidence level required for the test is selected, usually 95%.

   e.g., $Q_{crit} = 0.625$ (95% confidence level)

4. If $Q_{exp} > Q_{crit}$, then the suspect value can be characterised as an outlier.

   e.g., $Q_{exp} > Q_{crit}$, therefore data point 27.5 can be characterised as an outlier.

Table 21 Critical values of Q for Dixon's test, from Rorabacher[5]

n	Q_{crit} CL at 90%	Q_{crit} CL at 95%	Q_{crit} CL at 99%
3	0.941	0.970	0.994
4	0.765	0.829	0.926
5	0.642	0.710	0.821
6	0.560	0.625	0.740
7	0.507	0.568	0.680
8	0.468	0.526	0.634
9	0.437	0.493	0.598
10	0.412	0.466	0.568

The problems associated with the Q-test are that it may be misleading if more than one outlier is present, and also if an outlier is detected, a decision then has to be made about whether to exclude the data point from further statistical calculations. Methods which are more robust than the Q-test, such as the Huber method (described in the AMC technical brief, no. 5[6]), are becoming increasingly favoured for treatment of outliers since they consider all data present in the set, and not only three data points as in the Q-test. Robust statistics utilise approaches such as the use of the median and median absolute difference to estimate the mean and standard deviation respectively. In this way the outlying data has no effect and the does not have to be rejected. Any approach used to deal with outliers will have to be justified fully in the validation report.

Results for Each Validation Characteristic

The typical results which may be obtained for each validation characteristic are discussed below and the results which should be included in the validation report, as per ICH guidelines[1], are summarised in Table 23.

Specificity

The results obtained from specificity studies may take a number of forms. Qualitative data includes spectra from spectroscopic techniques, chromatograms for chromatographic techniques and graphical outputs from for example, differential scanning calorimetry (DSC). These types of results may be interpreted by a visual inspection to ensure that the spectra are as expected or that the peaks in the chromatograms are well resolved. Quantitative interpretation of this data may also be performed, the software used to manipulate spectral information often enables a comparison of spectra to obtain a score based on how well they match. The resolution of peaks may be calculated for chromatographic techniques to show that the peaks are adequately separated.

Quantitative data is obtained from specificity studies where the method is based on an absolute measurement, such as melting point. In an identification based on this type of method the measurement taken from authentic material containing the analyte is shown to be different that that obtained for material which does not contain the analyte. In the situation where specificity is demonstrated by comparing the results obtained for the method, in the presence of potential interferences, to the results obtained using a second well characterised method, then the results from the two methods are tested for equivalence. A student t-test may be used for this purpose.

Linearity

To investigate the calibration model the responses obtained for the linearity solutions are plotted against the concentrations of those solutions. An initial visual inspection will indicate if the method is likely to be linear. Regression analysis is performed and the regression statistics are inspected. An example of regression analysis including the interpretation of the statistics obtained is provided in Chapter 2, pages 29 to 32.

Range

The range is defined by the result obtained for the linearity, accuracy and precision investigations. The results from these studies should be suitably linear, accurate and precise to validate the range of the method.

Accuracy

The samples prepared for accuracy are analysed following the analytical method and the results obtained provide an estimated value of the assay, impurities, etc. for each sample. This data is expressed as a percentage of the known value in the sample to obtain the percent recovery for each individual sample. Typically, the mean, %RSD and confidence interval for the data is then calculated. Additionally, regression analysis may be performed on the estimated values versus the known values to further investigate the data. An example of accuracy data which was obtained for an assay method validation study, and is interpreted using the approaches above, is presented in Chapter 2, pages 39 to 40. If accuracy is determined by comparison to a second method, then the results from each method need to be compared. A Student's t-test may be appropriate for this comparison.

Note that for impurities it is important to ensure that it is clear how the impurities results are expressed, e.g., weight/weight or area percent, in all cases with respect to the major analyte.

Precision

Since precision is a measure of the spread of data, the standard deviation is used to assess the data. Typically, authentic samples are used for precision and so these will generate a data set comprised of analytical results for each sample as per method. These individual results are used to calculate the mean, standard deviation, %RSD and 95% confidence interval.

For intermediate precision or reproducibility, the overall precision is calculated for the whole data set to assess the spread of values when the analytical method is performed using a range of conditions. To compare the data sets obtained for each condition or to compare the results from different laboratories, comparative statistics such as Student t-tests may be used. Examples of precision data are provided in Chapter 2, system precision in Table 9, and repeatability, intermediate precision, and reproducibility in Table 11.

Detection Limit and Quantitation Limit

The results for the detection limit and quantitation limit investigations consist of visual data, such as chromatograms, on which the signal to noise may be measured, or quantitative data if the determination is based on the standard deviation of the response or the slope. Once the quantitation limit has been determined it is validated by the analysis of a suitable number of samples to show that it is accurate and precise. The necessary data is generated in the same way as previously detailed for accuracy and precision.

Robustness

The interpretation of the results obtained from a robustness study using an experimental design may be rather complex. If experimental design software[7,8] is used for the study then this will assist with the interpretation process. The necessary steps for a robustness study are detailed in Chapter 2, pages 60 to 64, together with useful references for further information.

Stability

The results obtained for solution stability evaluation typically consists of values obtained for stored solutions and either the values obtained during the original analysis or values for freshly prepared control samples. The interpretation of these results is to compare the results for the stored samples to those obtained for the original analysis or the control samples. The comparison may be performed by expressing the stored result as a percentage of the original or control result, or a statistical test such as the Student's t-test may be used.

Example

3 test solutions were used to evaluate stability under refrigerated conditions for 48 hours and the following results were obtained:

Table 22 Example of results obtained for solution stability

Sample	Original results (mg/tablet)	Reanalysis (mg/tablet)	Reanalysis expressed as a percentage of the original result (%)
1	24.7	24.8	100.4
2	25.4	25.2	99.2
3	24.9	25.1	100.8
Mean	25.0	25.0	
SD	0.294	0.170	

The acceptance criterion for the solution stability is that the reanalysis is within 2% of the original analysis. Therefore, it is concluded that the solutions are stable refrigerated for 48 hours. A paired Student's t-test was performed to compare the means obtained for each data set. No statistically significant difference was detected.

Results to include in the validation report

A summary of the results which should be included in the validation report is presented in Table 23.

Table 23 Results to include in the validation report for each validation characteristic

Validation Characteristic	Typically included in the validation report
Specificity	Comment on the specificity of the method. Appropriately labelled representative spectra, chromatograms or other graphical outputs should be presented. For chromatograms, peaks should be identified and retention times given. Quantitative data, as appropriate. Where compete discrimination is not achieved include comments relating to any other methods which are used, e.g., for identification more than one method may be used, and non-specific assay methods may be supported by results from an impurities determination.

Table 23 Results to include in the validation report for each validation characteristic

Validation Characteristic	Typically included in the validation report
Linearity	Comment on the linearity/calibration curve for the method. Plot of linearity data. Regression analysis statistical output: Equation of line Correlation coefficient Residual sum of squares 95% confidence interval for the intercept Plot of residuals
Range	The validated concentration range of the method.
Accuracy	Comment on the accuracy of the method. Express results as percent recovery and provide either individual and mean values or the difference between the mean and actual value, with confidence intervals. Results from comparison to second method, if appropriate.
Precision	Comment on the precision of the method. The individual results, the mean, standard deviation, %RSD and confidence interval are reported for each type of precision.
Detection Limit	Report the detection limit and present the method used for the determination. For a detection limit which is based on signal to noise or visual evaluation present relevant graphical data, e.g., chromatogram.
Quantitation Limit	Report the quantitation limit and present the method used for the determination. Present data related to the precision and accuracy of the quantitation limit, or refer to the accuracy and precision sections of the report. Present relevant graphical data if appropriate, e.g., chromatogram.
Robustness	Comment on the robustness of the method with respect to the factors investigated. Present the data obtained from the experimental design investigation and comment on the system suitability test used for the method.
Stability	Comment on the stability of the test solutions for the method. Present the data obtained for test solution stored at the studied conditions with appropriate comparison to demonstrate stability.

Non-compliance with acceptance criteria

In the event of results being obtained which do not meet the acceptance criteria an investigation should be performed to determine the cause. Redevelopment of the method may be required if it is concluded that it is not suitable for the intended purpose.

Documentation

Similar to the requirements for the validation protocol, the format and the approval of the documentation for the validation report will depend on the requirements of the in-house validation policy for a particular laboratory. In early phase studies tabular summaries of the results are commonly used, these provide all the information required for submissions to support clinical trial applications.

In late stage development the report is usually more extensive, containing all the results obtained during the study and a full discussion of those results including detailed discussion of each validation characteristic and also relevant information obtained during method development. The justifications for the methods used and the conclusions from the study should be detailed.

Revalidation

The life cycle of an analytical method was introduced in Chapter 1 and the diagram used to illustrate this cycle is reproduced in Figure 13. It can be seen from the figure that the validation of an analytical method is part of a cycle and if any changes to the method occur the effects need to be evaluated to determine if revalidation is required. Changes to the method may include changes to the material being tested such as different manufacturing routes or specification limits. Typical examples are provided in Table 24, also reproduced from Chapter 1.

Figure 13 The life cycle of an analytical method

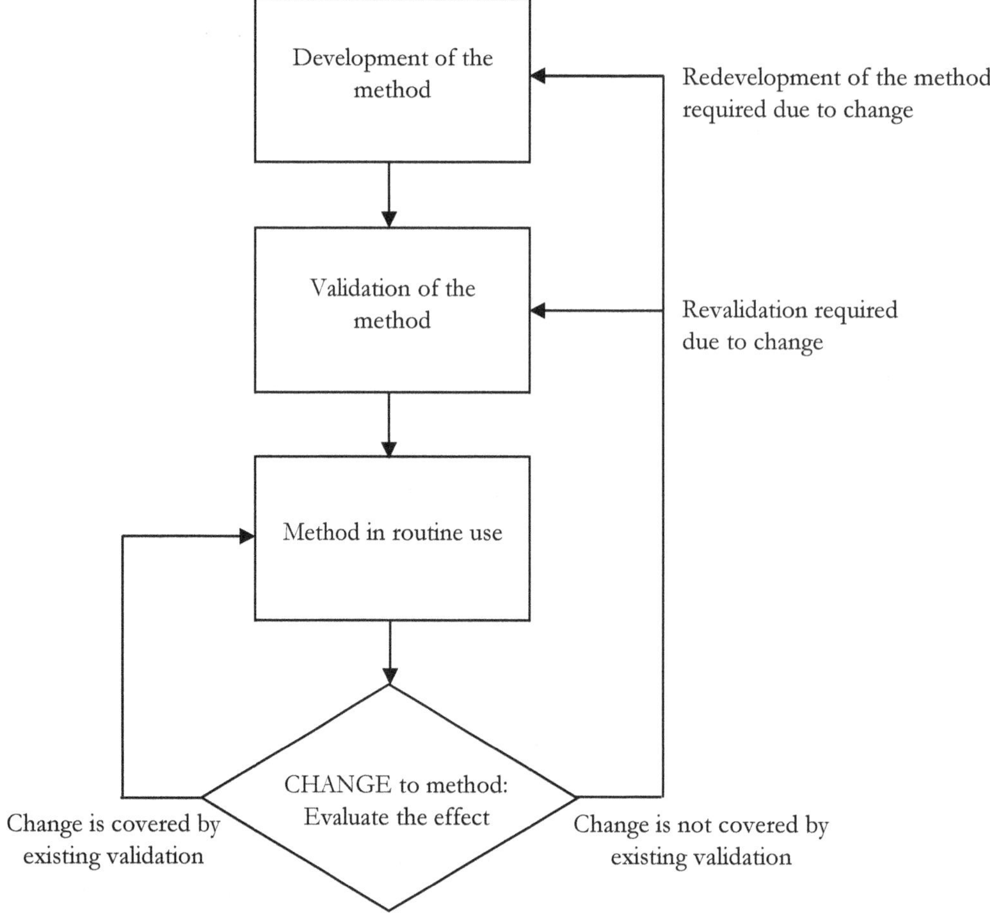

Table 24 Examples of potential changes to analytical methods

Source of Change	Change	Effect
Synthesis of the drug substance	Different synthetic route	Revalidation and potentially redevelopment of the method
Composition of the drug product	New excipient	Revalidation
	New dose size	Revalidation
Specification limits	Significantly lower impurity limits	Revalidation and potentially redevelopment of the method
	Significantly higher impurity limits within validation limits	None
New application of the method	New compound	Development and validation of new method

Case Study

A validation protocol for the analytical method presented in Chapter 3, 'Analytical Method for Determination of the Assay and Degradation Products of 'MiracleCure' 25 mg Tablets by HPLC' (Figure 12), was designed previously (refer to Table 19). The results from this study are presented in a tabular summary format in Table 25.

Table 25 Summary validation report for the analytical method and validation protocol detailed in Chapter 3 (refer to Figure 12 and Table 19)

Validation characteristic	Results	Acceptance criteria
Specificity	Chromatograms obtained for the specificity solutions showed that there are no interfering peaks. Retention times: MiracleCure 10.3 minutes Impurity X 12.4 minutes DP 1 15.0 minutes DP 2 37.2 minutes Excipients no peak detected Critical pair resolution = 3.2 (MiracleCure and Impurity X)	No peaks interfering with the peaks due to MiracleCure, Impurity X, DP 1 or DP 2. Minimum resolution between peaks of interest and neighbouring peaks = 2.
Linearity – Assay	Calibration model is valid Equation of line, y = 1219x + 691 Correlation coefficient, r = 1.000 Residual sum of squares = 150186 95% confidence interval for the intercept includes zero, -105 to 1487, intercept is 0.56% of nominal response. The plot of residuals shows random scatter and no systematic trends.	Calibration model valid. r > 0.999 Residual plot shows random scatter and no systematic trends. 95% confidence interval of intercept includes zero, or intercept is less than 2% of nominal response.
Linearity – DP 1	Calibration model is valid Equation of line, y = 1303x – 3.17 Correlation coefficient, r = 0.998 Residual sum of squares = 4318 95% confidence interval for the intercept includes zero, -18.47 to 12.12, intercept is 0.95% of response at 50% SL. The plot of residuals shows random scatter and no systematic trends.	Calibration model valid. r > 0.99 Residual plot shows random scatter and no systematic trends. 95% confidence interval of intercept includes zero, or intercept is less than 2% of response at 50% SL (0.25% nominal).

Table 25 Summary validation report for the analytical method and validation protocol detailed in Chapter 3 (refer to Figure 12 and Table 19)

Validation characteristic	Results	Acceptance criteria
Linearity – DP 2 & unknown degradation products	Calibration model is valid Equation of line, y = 1231x – 0.391 Correlation coefficient, r = 0.996 Residual sum of squares = 1096 The plot of residuals shows random scatter and no systematic trends. 95% confidence interval for the intercept includes zero, -9.66 to 8.88, intercept is 0.31% of response at 50% SL. Difference between slopes for assay and slope for impurities = 0.98%	Calibration model valid. r > 0.99 Residual plot shows random scatter and no systematic trends. 95% confidence interval of intercept includes zero, or intercept is less than 2% of response at 50% SL (0.1% nominal). Slope is within 5% of slope obtained for assay linearity.
Range	Range validated, refer to results for linearity, accuracy and precision.	Linearity, accuracy and precision demonstrated over the range of QL to 120% nominal concentration, or 0.0125 to 0.3 mg/mL.
Accuracy – Assay	Method is accurate for assay of MiracleCure over the required range. Mean recovery (n=9) = 99.9 SD = 0.58 %RSD = 0.58 95% CL = 99.5 to 100.3 Range of individual recoveries = 98.9 to 100.8 Regression analysis output: Slope = 1.005 95% confidence interval includes 1	Mean recovery within 98-102%. Individual recoveries within 97-103%. Regression analysis of known vs. estimated: Slope within 0.98-1.02 95% confidence interval of slope includes 1

Table 25 Summary validation report for the analytical method and validation protocol detailed in Chapter 3 (refer to Figure 12 and Table 19)

Validation characteristic	Results	Acceptance criteria
Accuracy – DP 1	Method is accurate for determination of DP 1 over the required range. Mean recovery (n=9) = 101.0 SD = 3.44 %RSD = 3.41 95% CL = 98.8 to 103.3 Range of individual recoveries = 95.9 to 105.8 Regression analysis output: Slope = 1.013 95% confidence interval includes 1	Mean recovery within 90-110%. Individual recoveries within 70-130%. Regression analysis of known vs. estimated: Slope within 0.9-1.1 95% confidence interval of slope includes 1
Precision	System precision, repeatability and intermediate precision all comply with the acceptance criteria and thus the method is precise over the required range.	
System Precision	System precision for 6 replicate injections of calibration standard = 0.35%	%RSD for 6 injections ≤ 1%
	System precision for 6 replicate injections of MiracleCure at QL = 2.96%	%RSD for 6 injections ≤ 10%
Repeatability – Assay	Results for 6 preparations from a representative batch of milled MiracleCure 25mg tablets. %RSD = 0.48% Mean = 100.0% label claim SD = 0.48 95% confidence interval = 99.7 to 100.4% label claim	%RSD for 6 assay results ≤ 2%

Table 25 Summary validation report for the analytical method and validation protocol detailed in Chapter 3 (refer to Figure 12 and Table 19)

Validation characteristic	Results	Acceptance criteria
Repeatability – DP 1	Results for 6 preparations from a stressed batch of milled MiracleCure 25mg tablets which contains DP 1 at a level of approximately 0.2% w/w w.r.t. MiracleCure. %RSD = 3.76% Mean = 0.204% w/w w.r.t. MiracleCure SD = 0.008 95% confidence interval = 0.198 to 0.210% w/w w.r.t. MiracleCure	%RSD for 6 DP 1 results ≤ 10%
Repeatability – DP 2	Results for 6 preparations from a stressed batch of milled MiracleCure 25mg tablets which contains DP 2 at a level of approximately 0.1% w/w w.r.t. MiracleCure. n =6 %RSD = 4.01% Mean = 0.101% w/w w.r.t. MiracleCure SD = 0.004 95% confidence interval = 0.097 to 0.104% w/w w.r.t. MiracleCure	%RSD for 6 DP 2 results ≤ 10%
Intermediate precision – Assay	Results from repeatability samples for Assay, analysed using experimental design. n = 24 %RSD = 0.68% Mean = 100.1 % label claim SD = 0.68 95% confidence interval = 99.8 to 100.3 % label claim	%RSD for 24 assay results ≤ 3%

Table 25 Summary validation report for the analytical method and validation protocol detailed in Chapter 3 (refer to Figure 12 and Table 19)

Validation characteristic	Results	Acceptance criteria
Intermediate precision – DP 1	Results from repeatability samples for DP 1, analysed using experimental design. n = 24 %RSD = 5.04% Mean =0.202% w/w w.r.t. MiracleCure SD = 0.01 95% confidence interval = 0.198 to 0.206% w/w w.r.t. MiracleCure	%RSD for 24 DP 1 results ≤ 15%
Intermediate precision – DP 2	Results from repeatability samples for DP 2, analysed using experimental design. n = 24 %RSD = 5.37% Mean = 0.100% w/w w.r.t. MiracleCure SD = 0.005 95% confidence interval = 0.097 to 0.102% w/w w.r.t. MiracleCure	%RSD for 24 DP 2 results ≤ 15%
Detection limit – DP 1	DL for DP 1 is 0.02% nominal. Peak due to DP 1 at 0.02% nominal (0.05 μg/mL) detected in all 6 samples, mean S/N = 3.9.	DL is 0.02% nominal S/N ~ 3
Detection limit – DP 2 and unknown degradation products	DL for MiracleCure is 0.02% nominal. Peak due to MiracleCure at 0.02% nominal (0.05 μg/mL) detected in all 6 samples, mean S/N = 3.3.	DL is 0.02% nominal S/N ~ 3
Quantitation limit – DP 1	QL for DP 1 is 0.05% nominal. Mean S/N = 11.2. %RSD (n=6) = 3.87% (Refer to results for accuracy)	QL is 0.05% nominal S/N ~ 10 %RSD for 6 DP 1 results ≤ 10%.
Quantitation limit – DP 2 and unknown degradation products	QL for MiracleCure is 0.05% nominal. Mean S/N = 10.5 %RSD (n=6) = 3.43% (Refer to results for accuracy)	QL is 0.05% nominal S/N ~ 10 %RSD for 6 MiracleCure results ≤ 10%.

Table 25 Summary validation report for the analytical method and validation protocol detailed in Chapter 3 (refer to Figure 12 and Table 19)

Validation characteristic	Results			Acceptance criteria
Robustness	No significant effects observed for any of the factors tested. System suitability test in method is satisfactory.			No factor studied has a significant effect on the assay results, or, factor effects are controlled by an appropriate system suitability test.
Stability	Test solutions are stable for 7 days at room temperature protected from light and refrigerated. % difference between original and reanalysis:			Results for MiracleCure in the original analysis and reanalysis within 2%. Results for DP 1 and DP 2 in the original analysis and reanalysis within 10%.
		7 days RT	7 days refrig.	
	MiracleCure	0.5	0.7	
	DP 1	3. 6	2.8	
	DP 2	4.2	3.1	
	RT = room temperature, refrig. = refrigerated			

Summary

1. The validation report details the results obtained for the analytical method validation study.

2. The statistics which may be required for validation includes: calculation of the mean, standard deviation, confidence intervals and relative standard deviation; regression analysis; comparative studies (Student's t-tests); and assessment of the significance of outliers.

3. The results obtained for each validation characteristic are calculated and interpreted with comparison to the acceptance criteria. Deviations should be justified.

4. Revalidation of the analytical method may be required due to changes in, for example, the manufacturing process for the drug substance.

References

1. International Conference on Harmonisation (ICH) of Technical Requirements for Registration of Pharmaceuticals for Human Use, Topic Q2 (R1): **Validation of Analytical Procedures: Text and Methodology**, 2005, www.ich.org.

2. Software 'Minitab', Minitab Inc., www.minitab.com.

3. Software 'Microsoft Excel', Microsoft Corporation, www.office.microsoft.com.

4. E. Mullins, '**Statistics for the Quality Control Chemistry Laboratory**', RSC, 2003, p135-183.

5. D. B. Rorabacher, *Anal. Chem*, **63**, 139-146, 1991, '**A Statistical treatment for rejection of deviant values: critical values of Dixon's "Q" parameter and related subrange ratios at the 95% confidence level**'.

6. Analytical Methods Committee, AMC Technical Brief, 2007, No. 6, '**Robust statistics: a method of coping with outliers**' (available on RSC website, www.rsc.org).

7. Software 'MODDE', Umetrics, www.umetrics.com.

8. Software 'The Unscrambler', CAMO Software, www.camo.com.

Further reading

P. C. Meier, R. E. Zünd, '**Statistical Methods in Analytical Chemistry**', Wiley, 2000.

J. Miller, J. C. Miller, '**Statistics and Chemometrics for Analytical** Chemistry', 5th ed., Prentice Hall, 2005.

J. Ermer, J. H. Miller, '**Method Validation in Pharmaceutical Analysis, a Guide to Best Practice**', Wiley, 2005.

D. M. Bliesner, '**Validating Chromatographic Methods: A practical guide**', Wiley, 2006.

Notes

Glossary

A

Acceptance criteria — The criteria which are applied to the results obtained from a validation study, if the results comply with the criteria, then it is concluded that the analytical method is fit for the intended purpose.

Accuracy — The closeness of agreement between a test result and the accepted reference value.

Active Pharmaceutical Ingredient (API) — The 'active' or the 'active pharmaceutical ingredient' is the substance in a drug preparation that is pharmaceutically active.

Analyte — The compound of interest to be analysed.

Analytical Instrument Qualification (AIQ) — Documented evidence that an analytical instrument performs suitably for its intended purpose and that it is properly maintained and calibrated.

Assay — An analytical method to analyse or quantify a substance in a sample.

B

Bioanalysis — The chemical analysis of biological samples, e.g., plasma, urine etc.

Bioassay — A biological test, measurement or analysis to determine whether compounds have the desired effect either in a living organism, outside an organism, or in an artificial environment.

Bioavailability — The amount of drug absorbed into the body.

Bioequivalence — The comparison of the expected in vivo biological equivalence of two proprietary preparations of a drug. If two products are said to be bioequivalent it means that they would be expected to be, for all intents and purposes, the same.

Biopharmaceutical — A drug produced by biotechnology.

Biotechnology — The application of scientific and engineering principles to the processing of materials by biological agents.

Buffer — A solution that maintains constant pH by resisting changes in pH from dilution or addition of small amounts of acids and bases.

C

Chiral — A compound that is asymmetric, it is not superimposable on its mirror image.

Chromatogram — A plot of detector signal output or sample concentration versus time or elution volume during the chromatographic process.

Chromatography — A technique for separating mixtures based on differential migration of components carried by a mobile phase through a stationary phase.

Chromatography data system — A software package which is designed to enable acquisition and processing of chromatography data.

Cleaning — Applied to pharmaceuticals, this is the process of cleaning equipment used in the manufacture of drug substance and formulated products so that the amount of residual drug is below an acceptable level.

Critical pair — The pair of components in a mixture which elute closest together and thus if the resolution of these two is acceptable then all other components will be adequately resolved.

D

Degradation products — Compounds which are formed due to the degradation of the drug molecule.

Design of experiments — See experimental design

Detection limit — The lowest amount of analyte in a sample which would be detected but not necessarily quantified as an exact value.

Diluent — A solvent used to dissolve a substance.

Dissolution testing — A standardised method for measuring the rate of drug release from a dosage form.

Drug product — A finished dosage form, for example, a tablet, capsule or solution that contains a drug substance, generally, but not necessarily, in association with one or more other ingredients.

Drug substance — An active ingredient that is intended to furnish pharmacological activity or other direct effect in the diagnosis, cure, mitigation, treatment or prevention of disease or to affect the structure or any function of the human body.

E

Efficacy — The effectiveness or ability of a drug to control or cure an illness.

Enantiomer — The mirror image forms of an asymmetric molecule having one asymmetric centre.

European Agency for the Evaluation of Medicinal Products (EMEA) — The European regulatory authority for testing and approval of drugs.

Excipients — The ingredients which are present in a drug product other than the drug substance.

Experimental design — An approach for organizing experimental work where a diverse and representative set of experiments are selected in which all factors are independent of each other despite being varied simultaneously.

F

Food and Drug Administration (FDA) — The drugs regulatory authority in the US.

Forced degradation — A study in which a sample containing the analyte is deliberately placed under stress conditions (e.g., humidity, thermal, etc.) in order to produce degradation products.

Formulated product — The drug molecule incorporated into a preparation which can be administered to patients, also known as the drug or medicinal product.

Formulation — The details of the excipients and drug substance which are combined to make a drug product.

G

Good Laboratory Practice — Scientific codes of practice that apply to a pharmaceutical company's research laboratories and which are

	monitored by regulatory authorities.
Good Manufacturing Practice	Scientific codes of practice that apply to a pharmaceutical company's production plants and which are monitored by regulatory authorities.

H

Hetereoscedasticity	In regression analysis, non-constant variance over the levels of the factor being studied.
High Performance Liquid Chromatography	The modern, fully instrumental form of liquid-phase chromatography technique that uses small particles and high pressures. Sometimes called 'high pressure' LC.

I

Immunoassay	A test that uses the binding of antibodies to antigens to identify and measure certain substances.
Impurities	Undesirable components in samples of drug substance or drug product.
Intermediate precision	A form of precision which expresses within-laboratories variations: different days, different analysts, different equipment, etc.

L

Ligand-binding assay	See immunoassay.
Limit of detection	See detection limit.
Limit of quantification	See quantitation limit.
Linearity	The ability of an analytical method (within a given range) to obtain test results which are directly proportional to the concentration (amount) of analyte in the sample.
Liquid Chromatography – Mass Spectrometry	The combination of the analytical techniques, HPLC and mass spectrometry.

M

Metabolite	Any substance produced by metabolism or by a metabolic process.
Mobile phase	The solvent used in chromatography which moves the solute through the stationary phase.

N

New Chemical Entity (NCE) — A novel molecular structure for a drug.

New Molecular Entity (NME) — A novel molecular structure for a drug of biological origin.

P

Peak purity — The purity of a chromatographic peak, i.e., whether the peak is due to the analyte of interest only or is made up of co-eluting peaks.

Pharmacokinetics — The study of the metabolism and action of drugs with particular emphasis on the time required for absorption, duration of action, distribution in the body and method excretion.

Pharmacology — The study of the properties of drugs and their effects on living organisms.

Pharmacopoeia — A book containing directions for the identification of samples and the preparation of compound medicines, and published by the authority of a government or a medical or pharmaceutical society.

Placebo — A preparation that contains no active drug but looks and tastes similar to the preparation of the actual drug.

Polymorphic form — See polymorphism. Different polymorphic forms of a drug can have different chemical and physical properties which can have a direct impact on the process-ability of drug substances and the quality/performance of drug products.

Polymorphism — The ability of a drug substance to exist as two or more crystalline phases that have different arrangements and/or conformations of the molecules in the crystal lattice.

Potency — The amount of drug required to achieve a defined biological effect.

Precision — Expresses the closeness of agreement (degree of scatter) between a series of measurements obtained from multiple sampling of the same homogeneous sample under the prescribed conditions.

Q

Qualitative — Analysis which merely determines the components in a sample without any regard to the quantity of each ingredient; contrasted with quantitative analysis.

Quality control check — Routinely used for bioanalytical methods - a sample is prepared using the sample matrix and adding known amounts of analyte. It is used to check the performance of the method each time it is performed. If the results obtained for the QC check samples are within acceptable limits it indicates that the method has performed acceptably and thus the results obtained for the study samples will also be acceptable.

Quantitative — Analysis involving the measurement of quantity of the component in a sample.

Quantification — The act of expressing the quantity of something.

Quantitation limit — The lowest amount of analyte in a sample which can be quantitatively determined with suitable precision and accuracy.

R

Regression analysis — A statistical technique used to find relationships between variables.

Relative response factor — A factor used to correct for the different detector response experienced for different analytes using some detection techniques (e.g., UV).

Relative retention time — The retention time of a component in a mixture expressed relative to another component (usually the drug molecule).

Repeatability — Expresses the precision under the same operating conditions over a short interval of time.

Reproducibility — A form of precision which expresses the variability between laboratories.

Resolution — The degree of separation of chromatographic peaks.

Retention time — The time between injection and the appearance of the peak maximum.

S

Selectivity	A reaction that can occur with other substances but exhibits a degree of preference for the substance of interest.
Specificity	The ability to assess unequivocally the analyte in the presence of components which may be expected to be present.
Stability indicating method	An analytical method which produces results that may be used to indicate the stability of the drug substance or drug product, e.g., i.e., impurities method which can detect and quantify degradation products.
Standard deviation	Standard deviation is a measure of the spread or dispersion of a set of data.
Stationary phase	The chromatographically retentive immobile phase involved in a chromatographic process.
System precision	A measure of the variability of the measurement performed as part of an analytical method.
System suitability	A test designed to measure the overall performance of an analytical system to ensure that it is operating as expected. It is carried out each time an analysis is performed.

T

Tailing	A peak in a chromatogram which experiences tailing will have an extended trailing edge.
Toxicology	The study of the adverse effects of chemical and physical agents on living organisms.
Trueness	The closeness of agreement between the average value obtained from a large series of test results and an accepted value.

U

Ultra high pressure liquid chromatography	HPLC where the small particle size used in the columns generates pressures of up to approximately 100 MPa (equivalent to approximately 1000 bar or 14500psi).

V

Validation protocol — A document which details the design of a validation study and thus defines what experiments will be performed and how the results will be assessed.

Validation report — A document which details the results obtained from a validation study.

Variance — The square of the standard deviation.

W

Weighting — Process of systematically increasing the value of a particular data element or elements so as to give that element more significance in the analysis or calculations.

List of Abbreviations

AIQ	Analytical Instrument Qualification
API	Active Pharmaceutical Ingredient
CDS	Chromatography Data System
COTS	Commercial Off The Shelf
DAD	Diode Array Detector
DL	Detection Limit
DOE	Design of Experiments
DQ	Design Qualification
EMEA	European Medicines Agency
EP	European Pharmacopoeia
FDA	Food and Drug Administration (US)
GC	Gas Chromatography
GFC	Gel Filtration Chromatography
GLP	Good Laboratory Practice
GMP	Good Manufacturing Practice
HPLC	High Performance Liquid Chromatography
ICH	International Conference on Harmonisation
IMPD	Investigational Medicinal Product Dossier
IND	Investigational New Drug
IQ	Installation Qualification

JP	Japanese Pharmacopoeia
LC	Liquid Chromatography
LCMS	Liquid Chromatography Mass Spectrometry
LLOQ	Lower Limit of Quantification
LOD	Limit of Detection
LOQ	Limit of Quantification
MAA	Marketing Authorisation Application
MHLW	Ministry of Health, Labour and Welfare (Japan)
MHRA	Medicines and Healthcare products Regulatory Agency (UK)
MS	Mass Spectrometry
NCE	New Chemical Entity
NDA	New Drug Application
NME	New Molecular Entity
NMR	Nuclear Magnetic Resonance
OQ	Operational Qualification
PDA	Photodiode Array
pH	Parts hydrogen (measure of acidity)
PQ	Performance Qualification
QA	Quality Assurance
QC	Quality Control
QL	Quantitation Limit
RRT	Relative Retention Time
RT	Retention Time
SFC	Supercritical-fluid Chromatography

SL	Specification Limit
SOP	Standard Operating Procedure
SST	System Suitability Test
TLC	Thin Layer Chromatography
UHPLC	Ultra High Pressure Liquid Chromatography
ULOQ	Upper Limit of Quantification
USP	United States Pharmacopeia
UV	Ultraviolet

Index

www.ingramcontent.com/pod-product-compliance
Lightning Source LLC
Chambersburg PA
CBHW082306210326
41598CB00028B/4459
* 9 7 8 0 9 5 6 1 5 2 8 1 7 *